THE

MATHEMATICAL

TOURIST

Snapshots of Modern Mathematics

# THE MATHEMATICAL TOURIST

### Snapshots of Modern Mathematics

**IVARS PETERSON**

W. H. FREEMAN AND COMPANY
NEW YORK

**Library of Congress Cataloging-in-Publication Data**

Peterson, Ivars. The mathematical tourist: snapshots of modern
mathematics / Ivars Peterson.
p. cm. Bibliography: p. Includes index ISBN 0-7167-1953-3
ISBN 0-7167-2064-7 (pbk.)
1. Mathematics — Popular works.
I. Title. QA93.P475 1988 510 — dc19 87-33078 CIP

Printed in the United States of America

4 5 6 7 8 9 0 FF 6 5 4 3 2 1 0 8 9

To our son, Eric Arnis,
born while this book was taking shape

# CONTENTS

Preface   xiii

## 1
## EXPLORATIONS   1

## 2
## PRIME PURSUITS   15

## 3

## 4

# 5
## ANTS IN LABYRINTHS 113

# 6
## THE DRAGONS OF CHAOS 143

## 7
### LIFE STORIES 175

## 8
### IN ABSTRACT TERRAIN 213

# A GALLERY OF MODERN MATHEMATICS

**C O L O R   P L A T E   1   Minimal Möbius**
Glue together the ends of a long strip of paper after giving one end a half twist. The result is a remarkable figure called a Möbius strip, which has only one side and one edge. The same basic form also lies at the heart of a certain infinite minimal surface that twists around to intersect itself.

**C O L O R   P L A T E   2   Smooth Moves**
Graceful contours sweep out from the core of a recently discovered example of an infinite minimal surface that doesn't intersect itself. This particular surface can be modeled on a thrice-pricked doughnut. Two of the doughnut's puncture points are stretched out to form the hourglass shapes, while the third becomes the figure's equatorial skirt.

**C O L O R   P L A T E   3   Peeling a Psychedelic Onion**
A sphere isn't the only closed surface that has the same nonzero curvature everywhere. A self-intersecting torus, which doesn't look at all like a sphere, has a similar property. Peeling away the outside of this toroidal onion reveals an intricate inner geometry.

**C O L O R   P L A T E   4   Sliced Doughnuts**
A sphere can be sliced into a sequence of parallel circles and a pair of points at opposite poles. Similarly, a four-dimensional hypersphere can be visualized as two linked circles and a succession of surrounding doughnut-shaped surfaces. Rotating this representation of a hypersphere produces a complex pattern of intertwined rings.

**C O L O R   P L A T E   5   Mountains of the Mind**
The jagged path of a random walk, in which each step up or down is determined by chance, can be thought of as a cross section through a rugged, mountainous landscape. Such cross sections, along with appropriate shading and colors, can be used in computer graphics to construct a surprisingly realistic rendition of a mountain scene.

**C O L O R   P L A T E   6   Fractal Ferns**
Black spleenwort ferns don't grow only in flower pots or in shady spots by streams. They can also be created, pixel by pixel, on a computer screen. All it takes is a special set of simple equations that somehow captures the fine details evident in a real fern.

**C O L O R   P L A T E   7   A Tree Grows in BASIC**
One way to grow a tree is to let particles float about randomly until they hit and stick to a previously generated particle cluster. One by one, the particles aggregate into a treelike shape. The computer version of this process is called diffusion-limited aggregation. The colors correspond to the order in which particles arrive at the tree.

**C O L O R   P L A T E   8   Branching to Infinity**
Start with a number. Run it through an equation that generates a new number, then repeat the operation on the answer, and so on. The result is a sequence of numbers that, depending on a control parameter built into the equation, may eventually settle on a single value, or cycle through a small group of values, or randomly hop from one value to another. For a given equation, a computer can plot which of these possibilities occurs for different parameter values.

## COLOR PLATE 9  Culture Clash
Magnifying a portion of the Mandelbrot set's boundary reveals a chaotic world of winding tendrils, ragged shapes, and miniature copies of the snowmanlike Mandelbrot figure. Different colors show the rates at which various points near the boundary escape to infinity.

## COLOR PLATE 10  The Ubiquitous Snowman
Surprisingly similar features show up when both simple and complicated functions are iterated. For example, the pimply, plump figure characteristic of the Mandelbrot set often makes surprise appearances in the middle of maps associated with many different functions.

## COLOR PLATE 11  Basins of Attraction
Newton's method is a venerable technique for finding the roots, or zeros, of an equation. The expression $z^4 - 1$, where $z$ is a complex number, has four roots. In the complex plane, each root is surrounded by a basin of attraction. Normally, Newton's method nudges any starting point selected from a particular basin toward the appropriate root. However, points near the complicated boundaries between basins may behave in unexpectedly chaotic ways.

## COLOR PLATE 12  Bursting into Chaos
The exponential function, which can be used to represent compounded growth, can sometimes explode into chaos when the value of a parameter reaches a critical value. Such a system may provide a useful model for physical phenomena, such as combustion, that involve rapid transitions between chaotic and stable states.

## COLOR PLATE 13  Grid Rules
Cellular automata may consist of lines of cells, where each color represents one of five possible values. The value of each cell is determined by a simple rule based on the values of its neighbors on the previous line. The top pattern is grown from a single, colored cell, whereas the lower pattern begins as a random mix of colored cells.

## COLOR PLATE 14  Snowflake Cells
Starting with a single red cell in the center of a lattice of small hexagonal cells, a computer program step by step generates a snowflakelike pattern. The successive layers of ice are shown as a sequence of colors, ranging from red to blue every time the number of layers doubles.

## COLOR PLATE 15  Probabilistic Drip
Mathematicians are experimenting with models that consist of simple rules governing the way in which one cell on a checkerboard grid interacts with its neighbors. In oriented percolation, which simulates the trickle of oil through sand, the probability of transmission from one cell to another establishes flow. Superimposing the percolation patterns for three different probabilities shows that increasing the probability increases the extent of penetration.

## COLOR PLATE 16  Crystal Tiles
A mosaic made up of fat and thin diamond shapes is one example of a Penrose tiling. This curious mathematical structure has a fivefold symmetry, as seen in the starlike figures that dot the pattern. Decorating each tile with colored bands and imposing the rule that tiles be joined only if bands of the same color run continuously across a boundary ensure that the tiles do not fall into a periodic pattern.

# PREFACE

"Curiouser and curiouser," cried Alice.
—Lewis Carroll, *Alice's Adventures in Wonderland*

Years ago, as a University of Toronto undergraduate majoring in physics and chemistry (with a heavy dose of mathematics on the side), there were times when I felt I had dropped down a rabbit hole into a bewildering land. More than once, I remember sitting in cavernous lecture halls, surrounded by dozing or fidgeting classmates, trying to figure out what was going on. The lecturer would be scrawling equation after equation across the blackboard and speaking in a puzzling language that sounded like English but somehow wasn't. Although I could pin a meaning to practically every word, I didn't seem to understand anything he said. I was as lost as an accidental tourist wandering in a very foreign country.

That disconcerting feeling hasn't altogether disappeared. Now, as a journalist writing about science and mathematics, I often have the same dismaying sense of being an outsider when I listen to researchers presenting papers at professional meetings. The words sound so familiar, but the language seems so foreign. I suppose my students, during the 8 years that I taught high school science and mathematics, often had the same feeling, despite my best efforts to convey a sense of the fun and excitement to be found in those subjects.

Professional mathematicians, in formal presentations and published papers, rarely show the human side of their work. Frequently missing among the rows of austere symbols and lines of dense prose and among the barely legible mathematical formulae marching across transparencies projected on a screen is the idea of what their work is all about—how and where their piece of the mathematical puzzle fits in, the sources of their ideas, their fountains of inspiration, and the images that carry them from one discovery to another.

To most outsiders, modern mathematics is unknown territory. Its borders are protected by dense thickets of technical terms; its landscapes are a mass of indecipherable equations and incomprehensible concepts. Few realize that the world of modern mathematics is rich with vivid images and provocative ideas.

Mathematics itself is changing. The field shows a renewed emphasis on applications, a return to concrete images, and an increasing role for mathematical experiments. These changes are making mathematics generally more accessible to outsiders than before.

Gazing into mathland, one may now catch the glint of a fractal tower poking through the mist or feel the subtle pull of a swirling strange attractor. Sometimes the air murmurs with fragments of wondrous tales about mathematicians tangling with knots, poking into higher dimensions, pursuing digital prey, playing with soap bubbles, or wandering in labyrinths.

In more than 6 years of reporting for *Science News*, I've had a chance to talk with a number of the mathematicians responsible for changing the look and style of modern mathematics. I've seen many wonderful graphic images of deep mathematical concepts. Some, such as the fractal images created by Heinz-Otto Peitgen and his colleagues, even show up in art galleries and tour the world. I've learned that mathematics is full of intriguing, unanswered questions and that new discoveries and modes of thinking keep pushing mathematics onward. This volume is meant as an informal introductory guidebook for travel in the world of modern mathematics. It highlights several particularly interesting and exciting places.

Mathematics is a vast enterprise. Thousands of pages of original research, spread across hundreds of journals, are published every year throughout the world. The United States has nearly 10,000 mathematicians, comparable to the numbers of physicists and chemists. The topics in this book represent only a small fraction of what goes on in the field—just a few snapshots taken during brief forays into the sometimes bizarre and often fascinating wonderland of modern mathematics.

The scholarly writings of mathematicians are, in general, liberally sprinkled with footnotes and references to antecedents. In this way, mathematicians credit earlier work on which the structure of mathematics is carefully built. Conjectures are often associated with certain individuals; proofs may have a long history of partial successes that fence in conjectures more and more tightly. To save the reader from having to go through long catalogs of names associated with particular ideas, I have chosen to concentrate on concepts that I believe are most important. Similarly, I have usually mentioned only the key mathematicians responsible for a given piece of research. Books and articles listed in the bibliography at the end of this book provide more complete histories and fill in many of the details.

Because I'm not a native of the land of mathematics, I've had to rely on a host of mathematicians to guide me through mathematical thickets. I've borrowed freely from the works of many mathematicians and scientists, gleaning ideas and examples from numerous lectures, papers, articles, and conversations. Lynn Arthur Steen, Ron

Graham, and Ian Stewart's articles in *Nature* and *The Mathematical Intelligencer* have been very helpful in pointing me toward what is new in mathematics. References that I found especially useful are also noted in the bibliography.

I'm grateful to former *Science News* editor Joel Greenberg for allowing me to wander in obscure mathematical pastures in search of stories about mathematics and mathematicians. Much of the material in this book has appeared in a somewhat different form in *Science News* over the last 6 years. Two of those journeys in search of mathematics stand out. One was to Brown University in Providence, Rhode Island, where Tom Banchoff, in 1984, organized a special conference honoring the hundredth anniversary of the publication of *Flatland*. Banchoff and a variety of other speakers suggested many of the provocative ideas that surface in Chapter 4. The second foray was to the University of Calgary for a meeting on recreational mathematics, where I met Jerry Lyons, a senior editor at W. H. Freeman, and mathematician Richard Guy, who organized the conference. After I completed the manuscript for this book, Guy was kind enough to read it and point out a number of errors. Any errors that remain, however, are my responsibility.

Furthermore, I wish to thank the following mathematicians and scientists for helping me by explaining ideas, providing illustrations, or supplying reference materials: Tom Banchoff, Michael Barnsley, Joan Birman, Harold Benzinger, Manuel Blum, Ernie Brickell, Scott Burns, John Cahn, Nicholas Cozzarelli, Jim Crutchfield, Bob Devaney, A. K. Dewdney, Persi Diaconis, Rick Durrett, Jim Fisher, Howland Fowler, Martin Gardner, Solomon Golomb, Branko Grünbaum, John Guckenheimer, David Hoffman, John Hubbard, Leo Kadanoff, Huseyin Koçak, David Laidlaw, Edward Lorenz, Benoit Mandelbrot, Paul Meakin, Ken Millett, Rick Norwood, Peter Oppenheimer, Norman Packard, Julian Palmore, Heinz-Otto Peitgen, Roger Penrose, Charles Peskin, Carl Pomerance, Paul Rapp, Peter Renz, Manfred Schroeder, Gus Simmons, Paul Steinhardt, Jean Taylor, Bill Thurston, Anthony Tromba, Hugh Williams, and Stephen Wolfram. My apologies to anyone I have inadvertently failed to include in the list.

It required a considerable team effort to transform a stack of manuscript pages into a finished book. I am grateful to my wife, Nancy, for encouraging me to keep at it and for many helpful suggestions. I greatly appreciate the efforts of Jerry Lyons, project editor Philip McCaffrey, and everyone else at W. H. Freeman who worked so hard to make my first venture into book writing a rewarding one.

**Ivars Peterson**

# 1

## EXPLORATIONS

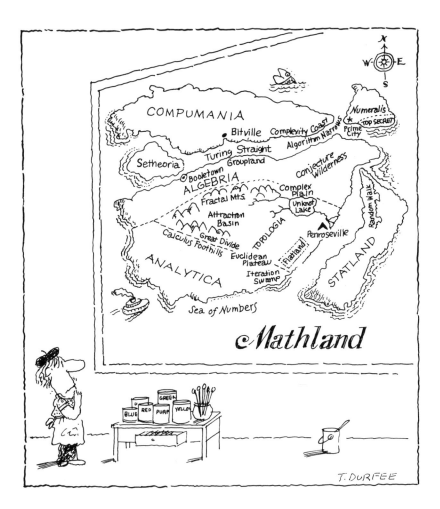

T. DURFEE

$A$ map is a picture of both the known and the unknown. Ancient mariners carried maps showing major cities and well-traveled trade routes alongside mysterious territories marked by fanciful names and decorated with mythical creatures. What wasn't known was guessed at. It was these unknown regions, with their promise of adventure and vague hints of fabulous wealth, that attracted early explorers.

A map of modern mathematics would reveal a similar mix of the familiar, the exotic, and the unknown. Algebra, trigonometry, and euclidean geometry, familiar to high-school students, lie in well-settled areas. Newer settlements, such as calculus, establish their spheres of influence nearby. Young upstarts — computer science, for one — nibble at old boundaries. Beyond the familiar are vast regions of mathematics still to be discovered.

## ——— MAPS OF A DIFFERENT COLOR ———

A deceptively simple mathematical problem lurks within the brightly colored maps showing the nations of Europe or the patchwork of states in the United States. It's the sort of problem that might worry frugal mapmakers who insist on decorating their maps with as few colors as possible. The question is whether four colors are always enough to fill in every conceivable map that can be drawn on a flat piece of paper so that no countries sharing a common boundary are the same color.

An additional definition and a condition turn this mapmakers' conundrum into a well-defined mathematical problem. A single shared point doesn't count as a shared border. Otherwise, a map whose countries are arranged like the wedges of a pie would need as many colors as there are countries. Moreover, countries must be connected regions; they can't have colonies scattered all over the map (see Figure 1.1, top left and middle).

First proposed in 1852 by British graduate student Francis Guthrie in a letter to his younger brother, the four-color problem has since intrigued and stumped professional and amateur mathematicians alike. Early on, mathematicians realized that three colors are certainly not enough for every possible map. It's easy to draw a map that needs four colors (see Figure 1.1, top right). English mathematician Augustus De Morgan also proved that it was impossible for five countries to be placed so that each one of them borders the other four. That led him to believe that five colors would never be needed. However, the proof that exactly four colors suffice remained elusive despite years of map sketching and attempted proofs.

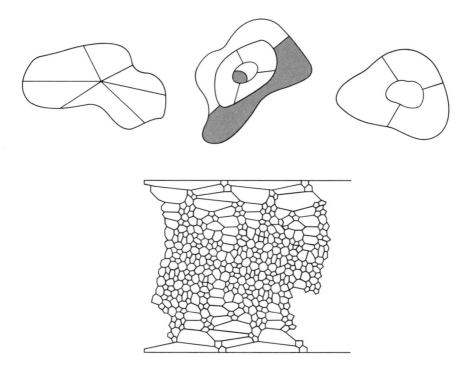

**FIGURE** 1.1    *Top left*, a single, shared point doesn't count as a shared border; *top middle*, countries can't have colonies; *top right*, why four colors are needed; *bottom*, try coloring this one.

In 1976, two mathematics professors finally proved the four-color theorem. It was the kind of news that mathematicians greet with champagne toasts. When Kenneth Appel and Wolfgang Haken of the University of Illinois announced their success, there were great celebrations in mathematical circles. They had scaled one of the Mount Everests of mathematics.

But there was a shock awaiting anyone curious about how the four-color theorem had been proved. The proof was unlike any other mathematical proof that had ever been done. Its complex text—strewn with diagrams and filling hundreds of pages—daunted all but the most determined readers. Even more dismaying to many mathematicians was that for the first time, a computer had been used as a sophisticated accountant to verify certain facts needed for the proof.

The computer allowed Haken and Appel to analyze a large collection of possible cases, a collection shown by mathematical analysis to be sufficient to prove the theorem. That task took 1,200 hours on a fast computer; it would have taken practically forever by hand. That also meant the proof could not be verified without the aid of a computer. Furthermore, Appel and Haken had tried various strategies in computer experiments to perfect several ideas that were essential for the proof.

Even now, doubts still linger about the validity of the proof of the four-color theorem. Although most mathematicians accept the idea that the Haken-Appel approach works, persistent rumors insist there is something wrong with the proof. Some people hint that the computer program was faulty; others, that the method itself is wrong in some way. A few complain that the computer program can't be verified properly. Nevertheless, the proof continues to hold up. Despite the discovery of such minor flaws as typographical and copying errors, the proof's basic strength has not been threatened.

To Appel and Haken, their proof of the four-color theorem represents a new and interesting facet of mathematics. Although a shorter, more elegant proof may someday be found, it's also possible that no such proof exists. The theorem may be one for which there never will be a proof short enough to be readily understood at a glance. It's an example of a curious occurrence in mathematics: the coupling of a simple, succinct problem with an incredibly complicated proof.

## LOGGING ON

More than a decade after Haken and Appel's proof of the four-color theorem, computers are still scarce among mathematicians and seldom used for serious mathematical research. One mathematician, Stanford University's Joseph Keller, even remarked in 1986 that at his university, the mathematics department has fewer computers than any other department, including French literature. Many mathematicians have the feeling that using a computer is akin to cheating and say that computation is merely an excuse for not thinking harder.

Still, computers are beginning to creep into mathematics. By providing vivid images that suggest new questions, they are helping to mend a rift that had developed between pure and applied mathematics. They are linking mathematical questions in computer science with computational questions in mathematics. These changes are enriching a subject that many outsiders have regarded as an abstract, even useless, pursuit.

The availability of very fast computers with large memories has led to two important developments in science and technology, both of which rely on mathematics. The first development is the use of mathematical equations to represent a physical situation, such as the air rushing over an airplane's wing or the chemical reactions leading to the formation of acid rain. When they are good enough, such mathematical models allow researchers to replace many of their wind-tunnel and test-tube experiments with computer manipulations.

The second development arises because computers are invaluable for extracting relationships and patterns hidden deep within vast amounts of data. For example, they can pinpoint the location of an oil field using seismic data or the site of a brain tumor using x-ray absorption measurements. The storage and processing of data require subtle, sophisticated techniques. More often than not, a piece of abstract mathematics worked out years before — and believed to be totally without practical value — finds a role in the "real" world.

One particularly striking example of the value of computer simulations and of the interaction between mathematics and science is in the work of applied mathematician Charles Peskin of New York University. He has spent more than a decade perfecting a computer model for blood flow in the heart. Peskin's aim is to create a tool that designers of mechanical heart valves can use to test prototypes without having to rely entirely on animal experiments and clinical trials with human patients.

Peskin has concentrated on modeling the heart's left side and on the movement of one particular valve, known as the mitral valve. With every heartbeat, the mitral valve's thin, flexible flaps of tissue smoothly slip out of the way when blood pushes forward. When the heart contracts, the valve snaps shut to keep blood from flowing back the wrong way. Peskin's mathematical model includes the characteristics of both the blood flow and the heart chamber's muscle tissue. Like an elastic band, the modeled tissue responds flexibly to the blood's pressure while it also exerts a force on the flowing blood.

In Peskin's heart model, the blood is represented by a large number of discrete points, each with a specific velocity and pressure. These velocities and pressures change as the fluid elements interact with their neighbors in ways that can be calculated using equations from physics. The heart's muscle fibers are modeled by a collection of moving particles joined by tiny springs. The properties of these elastic links change over time to simulate the change in tissue stiffness during a heartbeat.

Coupling flexible boundaries with a moving fluid made up of particles was perhaps one of the greatest mathematical challenges that came up during the development of this heart model. The mathematical and computational methods that Peskin and his collaborators developed to solve the problem also apply in many other situations —

from the movement of fish in water to the flow of suspended particles in a liquid.

The net result of the millions of computations required to make Peskin's model work is a sequence of pictures that can be strung together to produce a dramatic movie of a beating heart (see Figure 1.2). The vivid, two-dimensional images clearly show where flow is uneven, how differently shaped artificial valves respond, and even the likely performance of a weakened or diseased natural valve.

To an increasing number of practitioners, computer simulations represent a third way of doing science, alongside theory and experiment. In the past, a physical theory often consisted of a set of differential equations, which describe how a system changes over time or varies from place to place (as defined in Chapter 6). A theory today may consist of a computer program that models the way a system is supposed to evolve. Once experience shows that a given computer program accurately simulates a real system, then experiments may be done on the computer instead of in the lab. Of course, the computer model may produce results that fail to mimic reality. Then it's back to the drawing board.

The use of computer models is spreading rapidly to fields as diverse as astronomy, chemistry, and psychology and is illuminating the dynamics of pinwheel galaxies, the tremors that shake protein molecules, and the mechanisms that underlie human memory. In

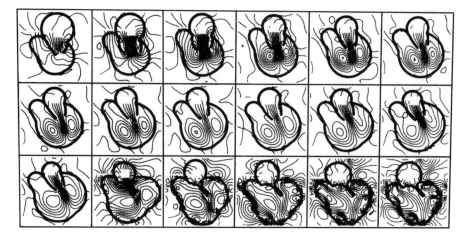

FIGURE 1.2    In this time sequence, computed streamlines show how blood flows through the heart's mitral valve.

many cases, mathematical models of physical situations save time and money. In aerodynamics, for example, typing in a few values at a computer terminal and seeing a graphics display of the results can save days otherwise spent crafting a wooden or metal model and testing it in a wind tunnel. Laboratory experiments and observations are still necessary, but now the researcher can choose which experiments are most likely to be useful and may even have a better idea of what behavior to observe and which variables to measure in a given experiment.

When an experiment produces masses of data, mathematical and statistical techniques are available to help researchers make sense of the jumble. The development of x-ray tomography and other techniques for probing the human body provides a good example of how much mathematics, both old and new, is needed to make such techniques work.

In computerized x-ray tomography, hundreds of x-ray beams, finer than needle points, zip through a slice of the human body. As each beam passes through tissue, bone, and blood, it weakens by an amount that depends on what it encounters. In mathematical terms, firing an x-ray beam through an object and seeing what happens to the beam is equivalent to finding "the projection of a function along a given line." This mathematical operation is called a Radon transform, named for a French mathematician who worked on such transforms early in the twentieth century. Radon's work appears to have been pure mathematics carried out for its own sake.

In the case of tomography, researchers must sift through their x-ray absorption data to compute the tissue density at a particular spot. In other words, from patterns hidden within all the x-ray intensity measurements, they need to work out the tissue arrangement that would give that particular set of intensities. That's not unlike looking at a complex pattern of ripples on a water surface and trying to decide where the ripples started and how many sources there were. Mathematically, the operation means computing the original set of values when only a set of averages of those values is known. It means performing a Radon transform in reverse.

Radon proved the basic theorems that specify under what conditions a function can be reconstructed from various projections or averages. Radon's ideas, however, apply only to continuous functions, which can be represented as smooth curves. In tomography, an infinite number of x-ray beams would have to be used to sample the entire cross section before the tissue densities at every spot could be accurately computed. Because only a finite number of x-ray measurements are actually made, mathematicians have had to work out approximate methods that allow cross sections of human organs to be reconstructed with a minimum of error. They try to make sure that an error isn't right where a tumor may be. New mathematical work is leading to faster algorithms for doing tomography computations so that x-ray images can be generated almost instantaneously.

Work on the discrete Radon transform and its inverse as applied to tomography has also suggested some interesting mathematical questions to investigate. Mathematician Ron Graham at AT&T Bell Laboratories and statistician Persi Diaconis at Stanford University are taking a close look at what can be deduced in situations in which certain averages are known but the original data are missing. Graham wonders, for example, to what extent secret data contained in confidential files can be uncovered if the right questions are asked.

Cracking a confidential data base may be something like the old parlor game of twenty questions. A player, receiving only yes-or-no answers yet asking the right sequence of pertinent questions, can often deduce the identity of some hidden object or person. In the same way, the answers to a series of general questions addressed to a particular data base could add up to a revealing portrait of something that is supposed to be secret.

A simple example shows how such a scheme might work. Suppose someone wants to find out Alice's salary. The inquisitor has access to information revealing that the average of Alice's and Bob's salary is $30,000; the average of Alice's and Charlie's salary is $32,000; and the average of Bob's and Charlie's salary is $22,000. This provides enough information to deduce that Alice's salary is $40,000.

Researchers often face a situation in which certain averages are known but the original data are missing. If eight data points happen to be identified with the eight vertices of a cube and each of the eight numbers is the average of its three nearest neighbors, then it's possible to deduce the actual but currently hidden value associated with each vertex *(see Figure 1.3)*. In this situation, the actual value at each vertex is equal to the sum of the nearest-neighbor averages minus double the average at the corner farthest from the point of interest. Curiously, the point that makes the largest contribution to the answer is the one that's farthest away.

Diaconis and Graham have developed a mathematical theory, based on the idea of discrete Radon transforms, that helps to decide how many and which averages are needed to crack a data base or to analyze statistical data. At the root of their exercise is the mathematical concept of how completely a bunch of averages captures the mathematical relationship underlying a data set.

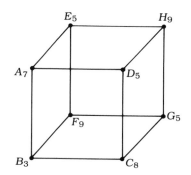

**FIGURE 1.3**  The actual value at vertex $A$ is the sum of the averages shown at $B$, $D$, and $E$ minus twice the average at $G$: $(3 + 5 + 5) - 2(5) = 3$. The values of the other vertices are 6, 3, 6, 9, 3, 12, 9.

## PURE POWER

To many people, mathematics—unchanging, reliable, dusty with age—has an aura of authority and rests on a firm foundation of pure logic. It promises certainty. Schoolchildren learn strict, seemingly infallible rules, ranging from the mechanics of addition to the intricacies of factoring algebraic expressions. Engineers routinely calculate specifications, counting on reference volumes filled with mathematical formulas. Stockbrokers use the logic built into elaborate computer programs to help make decisions about when to buy and sell.

But behind the apparently stolid, pristine, immutable public face of mathematics lies the exciting, turbulent, ever-changing world of mathematical research. Just as physics and other sciences go through episodes of both revolution and evolution, mathematics, too, changes and grows, not only in the way it is applied but also in its fundamental structure. New ideas are introduced; intriguing connections between old ideas are discovered. Chance observations and informed guesses develop into whole new fields of inquiry.

The territory of mathematics can be divided into three large chunks. The first, called algebra, involves the study of number systems. In general, an algebra consists of a number of mathematical entities (such as integers, matrices, vectors, or sets) and operations (such as addition or multiplication) with formal rules expressing the relationships between the mathematical entities. It includes, for example, the rules needed for adding, subtracting, multiplying, and dividing the ones and zeros (or binary digits) that zip through a computer's mind and memory.

Number systems and the operations that are performed within them can be classified in much the same way that animals can be divided into families and species. Just as zoologists can place cats in the family of mammals, mathematicians have places for algebraic systems that obey certain rules. One particularly important and useful category is the "group," which shows up in all branches of mathematics and in crystallography, particle physics, and other sciences.

Analysis, the second major piece of the mathematical continent, concerns functions. Functions express relationships. Simply put, a function is any rule that assigns a fixed output to a given input. For example, if the function is squaring, the number 3 is paired with the number 9, the number 7 with the number 49. The development of calculus, discovered independently in the seventeenth century by Isaac Newton and Gottfried Leibniz, is one of the central achievements of analysis.

The third region, geometry, is the study of the properties of shapes and spaces. Most people are familiar with the rigid forms of euclidean geometry — squares and cubes, circles and spheres, congruent triangles and parallel lines. But geometries can take on many different guises, reaching into higher dimensions and obeying various rules. Topology focuses on geometrical features that remain unchanged after twisting, stretching, and other deformations are imposed on a geometric space. Problems such as coloring maps, distinguishing knots, and classifying surfaces, or manifolds, not just in one, two, and three dimensions, but in higher ones as well, all fall within topology.

Lying a short distance offshore from the mathematical mainland are the islands of number theory and set theory. Number theory, at one time considered the purest of pure mathematics, is simply the study of whole numbers, including prime numbers. This abstract field, once a playground for a few mathematicians fascinated by the curious properties of numbers, now has considerable practical value. Identifying primes and finding the prime factors of a composite number play a crucial role in many cryptographic schemes — systems designed to keep secrets secret.

A set is any collection of entities, including numbers, that belong to a well-defined category. A "dog," for example, is a member, or element, of the set of all "four-legged animals." Numbers such as 57, −4, and 6,897 belong to the set of integers, whereas numbers such as 4.67 and $\pi$ do not. Set theory concerns the study of the structure and size of sets, as defined by various rules or axioms.

Somewhere near the mathematical continent lie the burgeoning landmasses of statistics and computer science. Both have close ties with mathematics, and the links are becoming increasingly important.

Computer science, for example, can be thought of as the study of *algorithms:* the methods or procedures used to solve given classes of problems. In cooking, a recipe is the algorithm that guides a cook in transforming a motley collection of ingredients into a scrumptious cake. Mathematicians also need recipes. In classical geometry, the ancient Greeks devised a slew of procedures, employing only ruler and compass as tools, for performing a variety of geometric feats, including the bisection of angles and the drawing of regular figures such as hexagons. Later mathematicians spent much of their time looking for algorithms for efficiently computing $\pi$, finding logarithms, identifying primes, and performing countless other mathematical tasks.

Today's explosive growth in computer use adds urgency to the investigation of algorithms. Because computers operate on the basis of a small, built-in set of operations, the programs that instruct them, which are essentially algorithms, must be made as efficient and reliable as possible. The mathematical analysis of algorithms has spawned the field called computational complexity.

Although mathematical methods play important roles in statistics and computer science, the focus in these fields is on accomplishing particular goals in an efficient, practical manner. In contrast, pure mathematics delves more deeply into the existence and nature of mathematical objects, even when these objects can't be computed or constructed.

Even what is traditionally called pure mathematics isn't immune to experiment and observation. To explore the properties of prime numbers, those special whole numbers divisible only by themselves and 1, mathematicians centuries ago compiled lengthy tables, using them to look for trends, to guess which properties such numbers have, and to see how primes are distributed among all whole numbers. The use of computers today merely facilitates this kind of list making to identify trends. Such computations, although not always an integral part of the final proof, often suggest what ought to be proved.

Numerical experiments, in which computers are used as untiring accountants and bookkeepers, have already suggested important ideas about the behavior of algebraic expressions and differential equations. One result of such experiments is the discovery of chaotic regions coexisting with islands of stability when certain dynamical

11

systems are simulated.' Out of this comes the disturbing news that some mathematical procedures—for instance, those used to solve equations—may not be as reliable as people had thought.

Experiments with soap films show the tremendous variability in the shapes of surfaces. Computer-generated pictures of four-dimensional forms reveal unusual geometric features. The crinkly edges of coastlines, the roughness of natural terrain, and the branching patterns of trees point to structures too convoluted to be described as one-, two-, or three-dimensional. Instead, mathematicians express the dimensions of these irregular objects as decimal fractions rather than whole numbers. Experiments and observations provide the heuristic hints vital for progress in mathematics.

But experiments and observations are not the whole story. Mathematics also involves proof. Once proposed, mathematical conjectures, or guesses, go through a trial by fire before they emerge as carefully defined theorems with a permanent place in the structure of mathematics. Mathematical ideas, once conceived, seem to have a life of their own and often wander far from their origins. For instance, the abstract concept of minimal surface, inspired by visions of soap films stretched across wire frames, now includes forms that would, as soap films, be too fragile or convoluted ever to be observed.

Mathematical research in modern times is an extensive enterprise. Because thousands of mathematicians publish hundreds of thousands of pages of new mathematical findings every year, most mathematicians find it difficult to keep up with what's happening across the field. Most of the time, they have to be content with being up-to-date in only a narrow furrow in the field. Consequently, mathematics tends to stay tightly packed into segregated compartments.

Nevertheless, many of the most striking mathematical results of the last decade are notions developed in one field that turn out to be a key element in solving outstanding problems in another, seemingly unrelated field. For example, in 1984, Dutch mathematician Hendrik Lenstra, out of curiosity, decided to study elliptic curves: equations of the form $y^2 = x^3 + ax + b$, where values for $a$ and $b$ are chosen arbitrarily. To his surprise, Lenstra serendipitously noticed a connection between elliptic curves and the age-old problem of determining the factors of an integer. His insight led to a new, speedier method of factoring large numbers. In another case of mathematical transfer, mathematician Vaughan F. R. Jones discovered a connection between operator algebras and knot theory, two topics that didn't have an obvious link. The result was an improved method for distinguishing different types of knots. British mathematician Simon Donaldson took the theory of Yang-Mills fields, which plays a role in the study of electromagnetic effects, out of physics and brought it to bear on a problem in topology. The result was the startling and unexpected discovery of a particularly bizarre geometry in four-dimensional space. All these examples attest to the essential unity of mathematics.

Even a brief glance at modern mathematical research reveals a dynamic enterprise of provocative ideas and questions. Far from being a domain of largely settled questions, mathematics is truly a wilderness. The well-mapped settlements lie few and far between, scattered across the continent and linked by a still skimpy network of highways and trails, some better traveled than others.

It's an exciting adventure to explore some of the newest paths that penetrate the mathematical wilderness — to pursue primes, to untangle twisted spaces, to delve into higher dimensions, to wander in labyrinths, to battle with chaos, and to puzzle over tiling patterns. These images may seem fantastic and metaphorical, but they're not. They are the objects studied by mathematicians — objects that we will see much more of on our tour. Novel landscapes, new vistas, and unexpected pathways lure the mathematical tourist to the frontier, where only questions and conjectures fill the scene.

# 2

## PRIME
## PURSUITS

A young child learning to count can get to 20 by counting fingers and toes. Higher numbers take on meaning as the child's experience grows. Mathematicians also find meaning in numbers. They start with patterns or relationships displayed by relatively small counting numbers and try to find out whether the patterns hold for ever larger ones. On such a foundation, they have built the discipline known as number theory.

In recent years, number theory has come out of its mathematical closet to play a crucial role in public affairs. The playful work of long-ago mathematicians turns out to have key applications in cryptography and computer security systems.

## HEADS OR TAILS?

Alice lives in New York City, and Bob lives in Los Angeles. They are recently divorced and communicate, when they do at all, only by telephone. Now they have to decide who should get an antique table that they have just jointly inherited. They agree to toss a coin. But neither one would like to choose, say, "heads" and then hear the other person exclaim over the telephone, "Here goes . . . I'm flipping the coin . . . you *lose!*" Somehow, Alice and Bob must find a way of tossing the coin without suspecting each other of cheating.

The dilemma confronting Alice and Bob is related to a problem that many computer users face. In the high-tech world of electronic mail and telephone-linked computers, methods are needed to guarantee that secret messages arrive intact, that funds are transferred correctly, and that contracts and agreements are properly signed. Otherwise, electronic letters are as open as postcards, and a person may be tempted to lie or cheat to gain an advantage over a rival.

One ingenious scheme for remote coin tosses blends the capabilities of modern, high-speed computers with some basic mathematics rooted in ancient Greece. In essence, this particular algorithm converts a call of "heads" or "tails" into the problem of factoring a large number. To win, the caller must find the two smaller numbers that the coin flipper multiplied together to generate the large number. This coin-toss algorithm not only provides Alice and Bob with a way to settle their dispute but also suggests protocols for ensuring the fairness of computer-aided business transactions, such as signing contracts and sending certified mail.

At the heart of the coin-toss algorithm is a procedure called the *oblivious transfer*, originally devised by Harvard University's Michael Rabin. The transfer is somewhat like a simple game played with a locked box requiring two different keys. A sender transfers the locked box to a recipient, who finds one key that partially unlocks the box. The sender has both keys and, without seeing what the recipient has done, must now pass on one of the two keys. Depending on which key is sent, the recipient will either succeed or fail in opening the box. Although the sender's choice controls the outcome, the sender never knows which choice to make to guarantee a certain result.

The oblivious transfer depends on two crucial mathematical ideas. First, mathematicians and computer scientists know several quick ways to determine whether a large number of, say, 60 digits, is a prime number, that is, evenly divisible only by itself and by the number 1. A properly programmed microcomputer can do this in a matter of seconds. Second, mathematicians believe that factoring a 120-digit or larger composite number, especially when the number is the product of just two roughly equal primes, can take years, even on the fastest available computers. The claim that factoring is intrinsically "hard" has not yet been proved, but mathematicians have assembled convincing evidence to support their belief. The security of Rabin's transfer hinges on the difficulty of factoring large numbers.

As in many modern cryptographic systems for sending secret messages, modular arithmetic plays an important role in the oblivious transfer. It's the kind of arithmetic that people use every day when they think about clocks and time. Adding 12 hours to the time shown on a clock face brings the clock's hands (or digits) right back to where they were at the start. Moreover, as far as the clock is concerned, adding 14, 26, or 38 hours is the same as adding just 2 hours.

In modular arithmetic, only remainders left over after division of one integer by another are saved. In the clock example, dividing 12 into 14, 26, or 38 produces a remainder of 2. Any other numbers that happen to come up during the computation are tossed out. More formally, given two integers, $a$ and $n$, the remainder of $a$ divided by $n$ is written as $a(\bmod n)$ and described as "$a$ modulo $n$." In this arithmetic system, $7(\bmod 5)$ is 2; $13(\bmod 5)$ is 3, and so on. A negative number, $-a$, is treated in the same way as $(n - a) \ (\bmod n)$. Thus, $-2(\bmod 5)$ is the same as $(5 - 2) \ (\bmod 5)$ or $3(\bmod 5)$, which equals 3.

Modular arithmetic is a handy key for locking up secrets. Its advantage is that it affords a kind of one-way trapdoor: what goes in doesn't easily get out again. Computing something like "14 modulo 12" is straightforward. The answer is 2. The reverse process is trickier. If the remainder happens to be 2 modulo 12 (or 2), then the original number could have been 2, 14, 26, or any one of innumerable other possibilities.

Here's how the oblivious transfer would work in the case of a remote coin toss. Alice and Bob use their linked personal computers to perform the toss. Normally, the computers would deal with 60- and 120-digit or larger numbers and probably go through the procedure in less time than it takes someone to read this example. However, to make sure that most of this chapter isn't taken up by lengthy strings of digits, Alice selects much smaller numbers than those a computer would handle.

Alice, in New York, starts the coin toss by selecting two prime numbers, 7 and 13. She multiplies the numbers together and sends the product, 91, to Bob. She keeps the numbers 7 and 13 secret. Bob wins the toss if he can factor 91, that is, find 7 and 13.

Suppose that Bob, in Los Angeles, can't factor 91. First, he can check whether Alice is cheating. She could have sent a number that can't be factored, but Bob's computer has efficient tests to show whether the number is a prime or a power of a prime. Once his computer finishes checking for primality, Bob randomly selects an integer, say 11, that falls between 1 and 91. The integer may, by chance, turn out to be a factor of 91, and therefore he wins the toss immediately. However, the chances of this happening with a 120-digit number are incredibly small. So Bob squares 11 to get 121. In the modular arithmetic step, he divides 121 by 91 and finds that the remainder is 30. Bob sends this remainder to Alice. He keeps the number 11 secret.

Alice knows the original number, 91, and she looks for all numbers less than 91 that generate a remainder of 30 when divided into her number. In this case, she can do it by trial and error, but mathematicians use a faster method based on the Chinese remainder theorem, which provides a handy way to reconstruct integers from certain remainders. She finds two pairs: $\pm 11$ and $\pm 24$. She can send either 11 or 24 to Bob. One of the numbers is Bob's number, but Alice doesn't know which one of the two is his.

If Alice sends the number 11, Bob would have no new information and would lose the toss because he wouldn't be able to factor 91. But Alice happens to send him 24. Bob adds his own number, 11, to 24 and gets 35. The greatest common divisor of 35 and 91 is 7. This number, 7, will automatically be a factor of 91. Interestingly, the greatest common divisor of 91 and the *difference* of 24 and 11 gives the other factor, 13. Now Bob can factor Alice's number, and he wins the coin toss. Finding the greatest common divisor is an application of Euclid's algorithm, one of the earliest known tools of number theory. The algorithm allows one to find, in a systematic way, the greatest common divisor of two integers without having to factor the two numbers.

The oblivious transfer also makes an appearance in a protocol for sending certified mail. The protocol was developed by Manuel Blum of the University of California at Berkeley. The usual problem with certified mail sent from one computer to another is that although senders get a record that the message was received, they don't get confirmation of the message's content. This allows the possibility of tampering and the appearance of deliberate or accidental errors in the received message.

In this case, the sender, Alice, embeds her message in the digits of 10 large numbers she sends to Bob for factoring. The numbers, each of which is a product of two large primes, are constructed so that Bob must factor all 10 in order to read the full message.

As in the coin-toss example, Bob picks a number at random and runs through the oblivious transfer. Depending on Alice's response, he may or may not be able to factor the first number. Then he goes on to the second number, runs through the algorithm, and so on, until he has tried all 10. After completing this first stage, Bob is likely to be able to factor some but not all of the numbers. In fact, because he has a 50 percent chance of factoring each number, he will, on the average end up with 5 of the 10 numbers needed. Bob then repeats the procedure, running through all of Alice's numbers in the same order as before but with new random guesses. He continues through these stages until he has enough information to find the factors of all 10 numbers. This usually takes about three or four passes through Alice's list.

Eventually, Bob has the information to find and decipher Alice's message, and Alice has a record of Bob's trial guesses. The collection of Bob's requests constitutes her receipt. If Bob were to deny receiving the message, a judge could determine from the transaction record that Alice provided enough information for Bob to factor the numbers and that Bob must have received the message she sent.

A simple variation on Blum's certified-mail scheme, with some additional safeguards, can ensure that contracts are signed simultaneously in different parts of the world. Traditionally, businessmen, diplomats, or generals have gathered in one place to sign contracts, agreements, or treaties. Because they don't trust each other, they want to see that the correct signatures are affixed at the appropriate time. Without a protocol and if the parties to a contract are in different places, someone may be tempted to cheat. All kinds of rituals and ceremonies have evolved over centuries to make sure that no person or country gains an advantage because of delays in signing a document.

A form of electronic signature is already familiar to people who use a bank's teller machines. To withdraw or deposit money, the customer must have a coded card and be able to enter a secret identification number. That combination constitutes the signing of a document permitting the transaction.

Blum's protocol is a more sophisticated and secure version of a bank's automated transaction system. Each person involved in the deal would send a copy of the contract to the other person, in the same way that each person would send certified mail by computer. Each person, then, also receives a receipt, which can be interpreted as a signature. The contract would include a clause stating that the contract is valid only if the participants have in their hands (or more likely stored in their computers) both contract copies and receipts. Only if the transaction is completed by both sides will the contract be valid.

We have described protocols of only two of many schemes proposed to reduce the risk of fraud when people use computers. Some schemes begin as simple games like mental poker or tossing a coin and evolve into elaborate protocols for electronic signatures or exchanging secret messages. These protocols often employ fundamental mathematical ideas that were discovered centuries ago without computers or applications in mind.

## ——— ——— PRIME PROPERTIES ———

The study of prime numbers has long been a central part of number theory, a field traditionally pursued for its own sake and for the beauty of its results. Once thought to be the purest of pure mathematics, this ancient pastime now figures prominently in modern computer science. The security of modern cryptosystems depends very strongly on the twin questions of how easy it is to identify primes and how hard it is to factor a large, random number. Neither question has a clear answer yet.

Divisible evenly only by themselves and the number 1, the primes stand at the center of number theory. Like chemical elements in chemistry or fundamental particles in physics, they are building blocks in the mathematics of whole numbers. All other whole numbers, known as composites, can be written as the product of smaller prime numbers. In fact, according to the fundamental theorem of arithmetic, each composite number has a unique set of prime factors. Hence, the composite number 20 can be broken down into the prime factors 2, 2, and 5. No other composite number has the same set of factors. The number 1 is considered to be neither prime nor composite.

Whereas chemists have only a hundred or so elements to play with, mathematicians interested in number theory must deal with an endless supply of primes. More than 2,000 years ago, Euclid proved that there is an infinite number of primes, and mathematicians ever since have been caught up in the never-ending pursuit of primes.

Euclid's argument is instructive. His proof relies on establishing a result that contradicts his initial assumption. The proof begins with the idea of a finite number of primes, with some largest prime, $p$. If all the primes are multiplied together and 1 is added to this gigantic product, a new number, $N$, emerges that is larger than $p$. If the initial assumption is correct, then $N$ can't be prime because otherwise it would be the largest prime. It must be a composite and divisible by at least one smaller prime number. However, because of the way $N$ was put together, all known primes, when divided into $N$, would leave a remainder of 1. Therefore, $N$ must be a prime that is larger than $p$. This contradicts the starting assumption and forces the conclusion that there can be no largest prime. There are, instead, infinitely many primes.

With such a vast array of numbers to work with, finding primes shouldn't be difficult. However, because prime numbers behave perversely and are scattered among the whole numbers in seemingly unpredictable ways, special nets are needed to catch them. Euclid's proof suggests one potential method for constructing prime numbers: multiply together all primes up to a certain value, then add 1. Is the result always a prime number? A simple experiment is revealing: $2 + 1 = 3$; $2 \times 3 + 1 = 7$; $2 \times 3 \times 5 + 1 = 31$; $2 \times 3 \times 5 \times 7 + 1 = 211$; $2 \times 3 \times 5 \times 7 \times 11 + 1 = 2,311$. It looks promising. But $2 \times 3 \times 5 \times 7 \times 11 \times 13 + 1 = 30,031 = 59 \times 509$. In fact, the next four instances prove to be composites. Further explorations show that Euclid's construction often but not always produces a prime.

In prime-number theory, simple, reasonable questions are remarkably easy to ask, yet many of these questions are surprisingly difficult or even impossible to answer. Is there a general formula that reliably generates prime numbers? Given a prime number, how far is it to the next one? Are there infinitely many twin primes, or consecutive pairs, such as 3 and 5, 41 and 43, and so on? No one knows yet. And there are dozens of similar conjectures waiting for proofs.

Fortunately, students of primes do have a systematic, though computer-consuming, way to trap primes and separate them from composites. Of course, they use a sieve for such a job. This particular sieve was discovered in the third century B.C. by Eratosthenes of Cyrene and goes by his name. The sieve of Eratosthenes generates a list of prime numbers by the process of elimination. To find all prime numbers less than, say, 100, the pursuer writes down all the integers from 2 to 100. First, 2 is circled, and all multiples of 2 (4, 6, 8, . . . ) are struck from the list. That eliminates all composite numbers that have 2 as a factor. This step, incidentally, also confirms that 2 is the only even prime number. A rather odd result, some would say.

The next unmarked number is 3. That number is circled, and all multiples of 3 are crossed out. The number 4 is already crossed out, and its multiples have also been eliminated. Five is the next unmarked integer, and the procedure continues in this way until only prime numbers are left on the list. In this example, the job finishes with 7 because 8, 9, and 10 are already gone, and 11 is greater than the square root of 100, the highest number in the table. All multiples of 11 less than 100 would already be crossed out. The sieve ends up trapping 25 numbers—all the primes less than 100 *(see Figure 2.1).*

| | | | | | |
|---|---|---|---|---|---|
| 1 | 2 | 3 | 4 | 5 | 6 |
| 7 | 8 | 9 | 10 | 11 | 12 |
| 13 | 14 | 15 | 16 | 17 | 18 |
| 19 | 20 | 21 | 22 | 23 | 24 |
| 25 | 26 | 27 | 28 | 29 | 30 |
| 31 | 32 | 33 | 34 | 35 | 36 |
| 37 | 38 | 39 | 40 | 41 | 42 |
| 43 | 44 | 45 | 46 | 47 | 48 |
| 49 | 50 | 51 | 52 | 53 | 54 |
| 55 | 56 | 57 | 58 | 59 | 60 |
| 61 | 62 | 63 | 64 | 65 | 66 |
| 67 | 68 | 69 | 70 | 71 | 72 |
| 73 | 74 | 75 | 76 | 77 | 78 |
| 79 | 80 | 81 | 82 | 83 | 84 |
| 85 | 86 | 87 | 88 | 89 | 90 |
| 91 | 92 | 93 | 94 | 95 | 96 |
| 97 | 98 | 99 | 100 | | |

**FIGURE 2.1** The sieve of Eratosthenes catches 25 primes among whole numbers less than 100.

Although some shortcuts simplify the sieving, it remains a tedious, time-consuming procedure, especially when the targets are dozens or hundreds of digits long. Sieving is a serious business for number theorists, and because they have no reliable formulas that automatically generate only prime numbers of any desired size, they use the sieve of Eratosthenes to fill in the gap. The claims of some mathematicians, both amateur and professional, to have found a magic formula for primes have invariably turned out to be false. In many cases, the formulas are merely cleverly disguised versions of the ancient sieve of Eratosthenes.

Because so many important ideas in number theory seem resistant to definitive analysis, experiment plays an important role in this field of mathematics research. Although their work differs from the experimental research associated with, say, test tubes and noxious chemicals, number theorists, like chemists and other researchers, often collect piles of data before they can begin to extract the principles that neatly account for their observations. Strict reliance on deduction—the hops, steps, and jumps from one theorem or logical truth to another that we usually associate with mathematics—isn't sufficient in number theory.

Few people outside of mathematics are aware of the field's empirical aspect. Much of the mathematics encountered by high-school and college students seems carved in stone, passed on unchanged from one generation to another. Yet even the fundamental principles of arithmetic and plane geometry were once the subjects of debate and speculation. It took centuries of constant questioning, brilliant guesses, and steady refinements to build the edifice now known as mathematics, and the structure continues to evolve. The study of prime numbers is one of the few areas still left in mathematics in which the concepts, questions and experiments are still simple enough to intrigue both amateur and professional mathematicians. In fields such as algebraic topology and differential geometry, usually only the professionals and advanced graduate students experience the thrill of the quest.

The search for patterns and trends plays an important role in the study of the distribution of prime numbers. Clearly, although Euclid proved that the list of primes goes on forever, the stretches between primes, on the average, get longer and longer (see Figure 2.2). The elimination of multiples of every new prime encountered by the sieve of Eratosthenes nicely demonstrates the increasing scarcity of primes among the larger integers.

At the same time, prime numbers appear to be scattered haphazardly throughout the whole numbers. In 1896, French mathematician Jacques Hadamard and Belgian Charles de la Vallée-Poussin

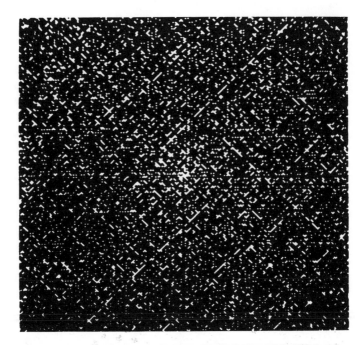

**FIGURE 2.2** A computer-generated grid showing primes (white dots) as a spiral of integers from 1 to about 65,000.

used this idea to find an expression that precisely describes the statistical distribution of primes. The theorem that they independently proved is now known as the prime number theorem. It proposes that the average distance between two consecutive primes near the number $n$ is close to the natural logarithm of $n$. When $n$ is close to 100, for instance, the natural logarithm of $n$ is approximately 4.6. In that range, every fifth number or so should be a prime.

But the prime number theorem doesn't eliminate the element of surprise. It doesn't specify exactly where a particular prime falls. That's left to the innumerable tests and experiments that make up the gathering-of-data phase of mathematics research. This data-collection operation now increasingly involves computers, which have extended the list of known primes well beyond the range of hand calculation.

Sometimes, tantalizing pattern fragments show up. The sequence 31, 331, 3331, 33331, 333331, 3333331, 33333331 looks promising: all the numbers in that sequence are primes. Alas, it fails in the next step. The number 333333331 is a composite with factors 17 and 19607843. Such accidental patterns lead nowhere.

Some simple equations can generate a surprising number of primes. The formula $x^2 + x + 41$, for example, produces an unbroken sequence of 40 primes, starting with $x = 0$ (see Figure 2.3). For values of $x$ up to 10,000, the formula produces primes almost half the time. However, beyond 10,000, the proportion of primes inevitably decreases. Another simple "prime-rich" formula, $x^2 + x + 17$, also discovered by the eighteenth-century Swiss mathematician Leonhard Euler, generates primes for all values of $x$ from 0 through 15. However, mathematicians have proved that no polynomial formula has only prime values, so a search for general formulas similar to the ones illustrated is fruitless.

Primes often appear as pairs of consecutive odd integers: 3 and 5; 11 and 13; 41 and 43; 179 and 181; 209,267 and 209,269; and so

| | | | | | | | | | | | | | | | | | | |
|---|---|---|---|---|---|---|---|---|---|---|---|---|---|---|---|---|---|---|
| 421 | | 419 | | | | | | | | 409 | | | | | | | | |
| | 347 | | | | | | | 337 | | | | | | 331 | | 401 | | |
| | | 281 | | | 277 | | | 271 | | 269 | | | | | | | | |
| | 349 | | 223 | | | | | | | | | 211 | | | | | | |
| | | 283 | | 173 | | | 167 | | | 163 | | | | | | | | |
| | | | | 131 | | 127 | | | | | | | 263 | | 397 | | | |
| | | | | | 97 | | | | | | | | | | | | | |
| | 353 | | 227 | | 71 | | | 67 | | 89 | | | | | | | | |
| | | | | | | 53 | | | | | | | | | | | | |
| | | | 229 | | | 73 | 43 | | | | 157 | | | | | | | |
| 431 | | | 179 | | 101 | | (41) | | | | | | | | | | | |
| | | | | 137 | | | | 47 | | | | 257 | | | | | | |
| 433 | | | 181 | | 103 | | 59 | | 61 | | | | | | | | | |
| | 359 | | 233 | | 139 | | 79 | | | 83 | | | | | 389 | | | |
| | | 293 | | | | 107 | | 109 | | | 113 | | 199 | | 317 | | | |
| | | | | | | | | | | 149 | | 151 | | | | | | |
| | | | | | | | 191 | | 193 | | | 197 | | | | | | |
| | | | | 239 | | 241 | | | | | | 251 | | | | | | |
| 439 | | | | | | | | | 307 | | | 311 | | 313 | | | | |
| | | 367 | | | | | 373 | | | | | 379 | | | 383 | | | |

**FIGURE 2.3** The primes between 41 and 439 plotted on a square spiral, beginning with 41 in the center. Values along the diagonal satisfy the formula $x^2 + x + 41$.

on. These twins are spread throughout the list of known prime numbers. Statistical evidence suggests that there are infinitely many twin primes, and mathematicians have estimated that in the range close to the number $n$, the average distance between one pair of twin primes and the next is close to the square of the natural logarithm of $n$. But no one has yet proved that there is an infinite number of twin primes; it remains one of the major unsolved problems in number theory. Over the last century, numerous mathematicians have attacked the problem in a variety of ways, but all have failed. The optimists believe that it may take another century of determined assaults before anyone succeeds.

At the same time, it's relatively easy to prove that consecutive primes can be as far apart as anyone would want. The sequence of numbers $n! + 2, n! + 3, n! + 4, \ldots n! + n$ shows this conjecture must be true. The expression $n!$ (read as $n$ factorial) is less a statement of mathematical excitement than a shorthand way of writing the product of all whole numbers from 1 to $n$. The number $n!$, where $n = 5$, for example, has a value of $1 \times 2 \times 3 \times 4 \times 5$, or 120. In the general case, $n! + 2$ is evenly divisible by 2, $n! + 3$ is evenly divisible by 3, and so on. Finally, $n! + n$ is evenly divisible by $n$. Therefore, all the numbers in the sequence are composite. The sequence can be made arbitrarily long by picking a sufficiently large number $n$.

All their suggestive properties indicate that primes are not randomly scattered among the integers, yet prime numbers, in their maddening perversity, defy all attempts aimed at putting them precisely in their places. To try to prove many conjectures about prime numbers, mathematicians therefore often must devise sneaky stratagems to get at their prey indirectly. They fence in the unruly conjecture, first staking out a wide swath of territory before gradually closing in on it. The pursuers then draw the net tighter, taking in corners and cutting down boundaries while scrupulously checking for loopholes. Finally, if all goes well, the conjecture is trapped.

One such attempt is the continuing work on the Goldbach conjecture, which views prime numbers as the building blocks of other numbers. The Prussian mathematician Christian Goldbach suggested the idea in a letter to Euler in 1742. His claim, expressed in modern terms, is that every even number greater than 4 is the sum of two odd primes (that is, primes other than 2). The number 32, for instance, is the sum of 13 and 19. This proposition is now known as the "strong" Goldbach conjecture. The "weak" Goldbach conjecture holds that every odd number greater than 7 can be expressed as the sum of three odd primes. In both cases, the same prime number may be used more than once in a single sum.

Neither conjecture has yet been proved, but over the years, there have been several near misses. In 1937, the Russian mathematician Ivan M. Vinogradov, a leading number theorist, showed that all "sufficiently large" odd numbers can be expressed as the sum of three primes. Although Vinogradov couldn't state explicitly what "large enough" meant, in 1956, his student K. V. Borodzin found that $3^{3^{15}}$ works as an upper bound. That number has more than 6 million digits. If all odd numbers smaller than $3^{3^{15}}$ are the sum of three primes, then the weak Goldbach conjecture is proved.

Computer searches have verified both the strong and weak conjectures up to at least a million. Mathematicians have also proved that if the weak conjecture is false, then there are only a finite number of exceptions. Thus the trap is closing in on the weak conjecture, and many mathematicians now regard it as "essentially" proved. The strong version is somewhat more elusive, but in 1966, Chinese mathematician Chen Jing-run showed that all large enough numbers can be expressed as the sum of a prime number and a number that is either prime or the product of two primes. The net tightens.

Despite such successes, some number theorists believe that certain central questions about prime numbers may have no solutions. In 1931, a startled mathematical world had to confront and learn to live with such a possibility, not just in number theory but everywhere in mathematics. Austrian mathematician and logician Kurt Gödel proved that any collection of axioms, such as those underlying euclidean geometry or the theory of infinite sets, leads to a mathematical system containing statements that can be neither proved nor disproved on the basis of those axioms. A theorem, then, can be undecidable: adding more axioms wouldn't help because some theorems would still slip through. Furthermore, Gödel's work implies that in some cases, mathematicians wouldn't even be able to decide whether a theorem can or cannot be proved. Uncertainty strikes in mathematics!

Various guesses about prime numbers, perhaps even the strong Goldbach conjecture, may truly turn out to be undecidable. Few mathematical fields other than number theory have so high a proportion of widely believed but unproved propositions. This doesn't mean that the conjectures are all likely to be false, but it implies that for some conjectures, an airtight case might not exist, despite the steady accumulation of persuasive evidence in its favor. Mathematical proofs require more than just overwhelming evidence.

Even so, the quest isn't hopeless. Near misses and proofs that confirm special cases or parts of conjectures hint that eventually techniques may be developed to handle the problems—although it could take centuries. That is far different from saying that a conjecture can never be proved, for a single breakthrough could put such suppositions as the number of twin primes suddenly within reach. Meanwhile, with the use of computers, specific knowledge about prime numbers is growing steadily, although considering the centuries of effort put in by numerous mathematicians, the theory of prime numbers still seems meager.

## HUNTING FOR BIG PRIMES

Success in the search for mammoth primes, which hide among composite numbers, requires a lot of patience, luck, and fortitude and takes some strategic planning using a few tools from number theory. The quarry are huge numbers such as $2^{44,497} - 1$. The question is whether this 13,395-digit mammoth, just one of an endless supply of candidates, is prime or composite.

Of course, big-prime hunters could stalk their game simply by writing out the number $N$ and dividing into it, one by one, all the integers, starting with 2. If no integers divide into the target number evenly, $N$ must be a prime. Going any further than the square root of $N$ during the search is pointless because factors are always found in pairs. If $N$ has a factor larger than the square root of $N$, then the factor's partner must be smaller. Using only odd numbers after division by 2 is another possible shortcut.

In the case of mammoth primes, however, the trial-division method proves to be hopelessly time-consuming. A computer able to carry out a billion trial divisions every second would take considerably longer than the current age of the universe to finish the job. Trial division works best for relatively small numbers, although even a 100-digit number is beyond reach unless it happens to have a small factor or a special structure.

The real trouble with trial division is that it provides more information than is required by the question of whether a number is prime. The method not only supplies the answer but also gives the factors of any number that happens to be composite. The idea, then, is to find techniques that pinpoint primality without at the same time finding factors. If a given number has no small factors, such techniques for testing primality are bound to be speedier than trial division and its variants.

In the seventeenth century, Pierre de Fermat provided the basis for primality-testing methods that don't depend on factoring. Today's faster algorithms are all descendants of what is now usually called Fermat's little theorem. One version of the theorem states that if $p$ is a prime number and $b$ any whole number, then $b^p - b$ is a multiple of

$p$. For example, if $p = 7$ and $b = 2$, the theorem correctly predicts that 7 divides evenly into $2^7 - 2$, or 126.

The theorem makes it possible to state properties of numbers that are too large even to be written out in decimal form. Knowing that $2^{44,497} - 1$ is a prime number (established in 1979), allows us to say that an unimaginably gigantic number such as $3^{(2^{44,497}-1)} - 3$ is evenly divisible by $2^{44,497} - 1$. No physically conceivable computer could ever handle such an enormous number, but mathematicians can still deduce some of its properties.

Fermat's little theorem can easily be transformed into a test for prime numbers. If $p$ is a prime number and $b$ is any whole number, then the remainder when $b^p - b$ is calculated and divided by $p$ will be zero. Thus, for $p = 11$ and $b = 2$, $2^{11} - 2 = 2,048 - 2 = 2,046$, and 2,046 divided by 11 has a remainder of zero. This works for all prime numbers. If the remainder is not zero, then the number is composite. The difficulty, however, is that some composite numbers also give a remainder of zero. These composites are called pseudoprime numbers. The number $2^{341} - 2$, for instance, is a multiple of 341, even though 341 is a composite, being the product of 11 and 31.

Although calculating the value of 2 (or some other number) to a high power seems more formidable than trial division, mathematical shortcuts are possible because only the remainder is of interest. Here, modular arithmetic enters the picture, by making it relatively straightforward to find the remainder when, for example, $3^{1,037} - 3$ is divided by 1,037. There's no need to calculate the value of $3^{1,037}$ itself. It turns out that $3^{1,037}$ is congruent to (or gives the same remainder as) $845 \pmod{1,037}$, and $3^{1,037} - 3$ is congruent to $842 \pmod{1,037}$. Therefore, the remainder on division by 1,037 can't be zero. The number 1,037 is composite. This quick test provides practically no clues about the factors of the number.

All that we need to turn the basic Fermat method into an efficient primality test is a way to weed out the pseudoprimes. Fortunately, pseudoprimes are rare, although for a given value of $b$, there are an infinite number of them. Below $10^{10}$, for example, there are 455,052,512 primes, but for $b = 2$, only 14,884 pseudoprimes exist over the same range of whole numbers. The composite number 561 is the smallest pseudoprime for all choices of $b$.

Because the basic Fermat test does an excellent job, mathematicians and computer scientists have sought to generalize Fermat's test to exclude fake primes. The idea is to do some additional tests to find out more about the composite number that may be masquerading as a prime number. Such tests not only would say "pass" or "fail" but also would provide extra information about the numbers being tested. Moreover, the tests should work on numbers that don't necessarily have a special form.

The results of the tests on a particular number resemble a mosaic, with information about the number caught up in its fragments. If we look at a sufficiently large piece of the mosaic, we get enough information to decide whether a number is a prime or a composite. Like a hologram, the nature of the number being tested can be found in any section of the mosaic, which helps cut down the amount of computer time needed to run this type of primality-testing algorithm. Usually, the tests provide information to conclude that any divisor of the target number must be among a very small set of numbers. If none of the divisors works, the number is a prime.

An especially efficient form of such an algorithm for testing any number for primality was developed in the early 1980s. Len Adleman, Robert Rumely, and Carl Pomerance all contributed toward the invention of the testing algorithm. Hendrik Lenstra then discovered several variations of the new method, which made the procedure easier to run on a computer. Lenstra and Henri Cohen twisted the algorithm into shape to make it more compatible with real computer applications. Tests of their elegant implementation show that arbitrary, 100-digit numbers can be checked in seconds. Previously used algorithms would probably have taken in excess of a 100 years to do the same job.

A crucial question in computer science concerns how much such an algorithm for primality testing can be speeded up. The framework for answering such questions lies in the theory of computational complexity, which classifies some computational problems as "easy" and others as "hard," depending on how much time a computer may take to work out a given problem in the worst possible case.

Let's illustrate that classification scheme by considering methods for testing the primality of any number with $n$ digits. If the number of computational steps grows at a rate of $kn^a$, where $a$ and $k$ are fixed and determined by the algorithm characteristics, then primality testing can be done in "polynomial time" and is considered "easy." In other words, if the time required for a computer to solve a problem shows polynomial growth according to, say, the cube of the number's length, then increasing the number of digits from 20 to 50 increases computing time from, perhaps, 1 second to 12 seconds. But if the number of steps grows at a rate of $ka^n$, then the problem is "hard": it can't be decided in polynomial time and needs "exponential time." An algorithm that performs exponentially, at, say, $3^n$ can turn a 1-second solution into a thousand-century nightmare with a modest increase in the number of digits. It's like the difference between racing along at a steadily increasing pace versus running at a rate that climbs higher and higher as the speed increases.

The new primality-testing method requires exponential time, but the exponent changes so slowly that for reasonably sized numbers, the algorithm behaves a lot like one that operates in polynomial time. The question of whether a polynomial-time algorithm will ever be found for primality testing is still open; so is the question of whether factoring is inherently a "hard" problem. Mathematicians and computer scientists tend to assume that primality testing is easy and factoring hard, but to date, neither assumption has been proved.

The determined pursuit of primes can sometimes lead into novel territory. Primality testing turns out to be incredibly easy if the hunter doesn't mind making a mistake once in a long while, a result that is good enough for many practical applications but not for mathematical proof.

The approach of allowing a tiny chance of error stems from the stunning discovery that playing the odds can be far more efficient than following a preset algorithm. Making a sequence of random guesses yields the right answer most of the time, and the longer you keep up the guessing, the better your chances of ending up with a certified prime. This guessing game hinges on the existence of some numerical property that composites usually possess and that primes never have. Suppose, for instance, that for a composite number $n$, at least three-quarters of the numbers between 1 and $n$ have a particular property that can be checked very quickly. The test simply consists of picking, say, 50 numbers and checking for the appropriate feature. If any have it, then $n$ can't be a prime. If all trial numbers pass the test, then $n$ is almost certainly a prime. In that case, the chance of $n$ being composite would be at most $(\frac{1}{4})^{50}$, or $10^{-30}$. If those odds aren't good enough, then trying another 50 numbers takes only seconds, and the chance of error becomes even tinier. Algorithms that depend on such random procedures provide confident guesses rather than firm answers about whether a large integer is prime. They represent a trade-off between greater speed and increased uncertainty.

Primality testing is also easy when the numbers involved have a special form. The largest known primes are found among *Mersenne numbers*, named for Marin Mersenne, the French abbot who studied them during the early part of the seventeenth century. Mersenne primes are as hard to miss as gaily striped circus elephants gamboling in an open field. They have difficulty hiding among the composites because their special structure allows the use of relatively simple tests to determine their primality.

A Mersenne number, $M_p$, has the form $2^p - 1$, where $p$ is a prime. If $M_p$ itself is a prime, then it is called a Mersenne prime. For example, $M_7 = 2^7 - 1 = 127$, which happens to be a prime. In 1644, Mersenne found that $M_p$ is a prime for $p = 2, 3, 5, 7, 13, 17$, and 19. $M_p$ is not prime for 11: $M_{11} = 2^{11} - 1 = 2{,}047 = 23 \times 89$. Based on the trends that he saw, Mersenne predicted that $M_p$ would be prime for $p = 31, 67, 127$, and 257, and he conjectured that no other such primes occur in that range.

Fermat, and later Euler, discovered methods that simplified the task of testing for primality in the case of Mersenne numbers. They proved that all factors of any $M_p$ must be of the form $2kp + 1$ and, at the same time, of the form $8n \pm 1$, where $k$ and $n$ are integers. For example, choosing $k = 1$ and 4, and $n = 3$ and 11, $2{,}047 = 23 \times 89 = (2 \times 1 \times 11 + 1) \ (2 \times 4 \times 11 + 1) = (8 \times 3 - 1) \ (8 \times 11 + 1)$. This discovery greatly reduces the number of possible factors of $M_p$, making it easier to test for primality. It allowed Euler to show that $M_{31} = 2{,}147{,}483{,}647$ is indeed a prime.

In 1876, Edouard Lucas came up with a primality test suitable for all numbers, which also turned out to be particularly well tailored for testing Mersenne numbers. This test quickly showed that Mersenne's conjecture was incorrect. In 1930, D. H. Lehmer improved the Lucas method to form the basis for the test still used today.

The algorithm for testing the primality of Mersenne numbers begins by setting the initial value of a function $u$ equal to 4. A formula relates each successive new value of the function to the previous old value. That is, the new value $u\ (i + 1)$ is equal to the remainder after the old value $u(i)$ is squared, then decreased by 2 and divided by the Mersenne number itself. In mathematical terms, this is written as $u(i + 1) = (u(i)^2 - 2)(\mathrm{mod}\ M_p)$. Modular arithmetic strikes again! $M_p$ is a prime if after going through this procedure up to but not including the value of $p$, the final remainder is zero.

For example, if $p = 5$, $M_5 = 2^5 - 1 = 31$; $u(1) = 4$; $u(2) = (4^2 - 2)(\mathrm{mod}\ 31) = 14$; $u(3) = (14^2 - 2)(\mathrm{mod}\ 31) = 8$; $u(4) = (8^2 - 2)(\mathrm{mod}\ 31) = 0$. $M_5$ is a prime!

With the aid of high-speed computers, the Lucas-Lehmer test has turned out to be a relatively quick and easy way to test the primality of Mersenne numbers. In 1978, two California high-school students, Laura Nickel and Curt Noll, used 440 hours on a large computer to set the record for the highest known prime of that time. Their number, $2^{21,701} - 1$ was the twenty-fifth Mersenne prime discovered and has 6,533 decimal digits.

Thereafter, the pursuit quickly became a private game for the largest and fastest supercomputers. The historical record shows that it takes four times as much computation to discover the next Mersenne prime as it would to rediscover all previously known Mersenne primes. In a sense, the search for Mersenne primes can be seen as a

| Value of p for which $2^p - 1$ is prime | $2^p - 1$ | When Proved Prime | Machine Used |
|---|---|---|---|
| 2 | 3 | | |
| 3 | 7 | Antiquity | |
| 5 | 31 | | |
| 7 | 127 | | |
| 13 | 8,191 | 1461 | |
| 17 | 131,071 | 1588 | |
| 19 | 524,287 | | |
| 31 | 2,147,483,647 | 1772 | |
| 61 | 19 digits | 1883 | |
| 89 | 27 digits | 1911 | |
| 107 | 33 digits | 1914 | |
| 127 | 39 digits | 1876–1914 | |
| 521 | 157 digits | | |
| 607 | 183 digits | | |
| 1,279 | 386 digits | 1952 | SWAC |
| 2,203 | 664 digits | | |
| 2,281 | 687 digits | | |
| 3,217 | 969 digits | 1957 | BESK |
| 4,253 | 1,281 digits | 1961 | IBM-7090 |
| 4,423 | 1,332 digits | | |
| 9,689 | 2,917 digits | | |
| 9,941 | 2,993 digits | 1963 | ILLIAC-II |
| 11,213 | 3,376 digits | | |
| 19,937 | 6,002 digits | 1971 | IBM 360/91 |
| 21,701 | 6,533 digits | 1978 | CDC-CYBER-174 |
| 23,209 | 6,987 digits | 1979 | CDC-CYBER-174 |
| 44,497 | 13,395 digits | 1979 | CRAY-1 |
| 86,243 | 25,962 digits | 1983 | CRAY-1 |
| 132,049 | 39,751 digits | 1983 | CRAY X-MP |
| 216,091 | 65,050 digits | 1985 | CRAY X-MP/24 |

**F I G U R E   2.4**   Since 1952, computers have taken over in the search for Mersenne primes. In 1988, Walter Colquitt and Luther Welsh, using an NEC SX-2 supercomputer, discovered the thirty-first Mersenne prime, for which $p = 110,503$.

measure of increases in computing power over the last two centuries.

One recent champion, the record holder of 1985, was found accidentally while a new supercomputer was being put through its paces to make sure the machine was functioning properly. The number, the thirtieth known Mersenne prime, has an exponent of 216,091 and runs to 65,050 decimal digits. The computer took about three hours to complete the 1.5 trillion calculations involved (see Figure 2.4).

Where will the next mammoth prime be found? Are there an infinite number of Mersenne primes? No one knows yet.

## —————— BREAKING UP IS HARD TO DO ——————

A decade or so ago, an interest in factoring was the mark of an eccentric. Only a handful of number theorists cared to wrestle with lengthy strings of digits. This small, obscure group of mathematicians worked quietly, prying open large composite numbers to unlock their prime secrets. They reveled in the pure delight of calculation and in the pleasure of devising elegant algorithms to do their work. Like many ardent hunters, they even kept a list of "wanted" and "most wanted" targets (see Figure 2.5).

Those mathematicians are still around, but they have been joined by a crowd of computer scientists and applied mathematicians eager to test their skills. Armed with advanced algorithms, sophisticated computers, and specialized calculating machines, the newcomers are rapidly changing the formerly sedate and sheltered world of factoring. In just a few years, successful factorizations of "hard" numbers have jumped from numbers with 50 decimal digits to those with more than 80. Now, the list of "most wanted" factorizations must be updated regularly.

Factoring is of considerable interest because the security of several important cryptographic systems depends on the difficulty of factoring numbers with 100 or more digits. Without doubt, compared with testing for primality, factoring is hard. It takes a day or so of supercomputer time to break an 80-digit number that happens to have no small factors. It would take only a few seconds, if that, to tell whether the number is a prime. Much current research is aimed at providing as sharp an estimate as possible of how hard factoring really is.

The most obvious, systematic way to find the factors of a number $N$ is by trial division. If 2 doesn't divide evenly into $N$, then 3 may, or 4, or 5, and so on. The first number that divides evenly into $N$ is a

## MOST WANTED

| Number | Number of digits in "hard" part |
|---|---|
| $2^{211} - 1$ | 60 |
| $2^{251} - 1$ | 69 |
| $2^{212} + 1$ | 54 |
| $10^{64} + 1$ | 55 |
| $10^{67} - 1$ | 61 |
| $10^{71} - 1$ | 71 |
| $3^{124} + 1$ | 58 |
| $3^{128} + 1$ | 53 |
| $11^{64} + 1$ | 67 |
| $5^{79} - 1$ | 55 |

## FACTORIZATIONS

**FIGURE 2.5** The 1983 list of "most wanted" factorizations. Some of these numbers have a small, easy-to-find divisor, but that still leaves a "hard" part. A specially programmed supercomputer was able to crack all these numbers in the space of a year.

prime factor. Dividing $N$ by that factor and then starting the process all over again eventually leads to the full "prime decomposition" of $N$. To speed up the process, we can divide only by primes rather than by every integer up to $N - 1$; furthermore, we need not use trial divisors beyond the square root of $N$. This enhanced trial-division algorithm, as fine-tuned by computer scientists, works well for numbers that have 10 digits or so. Trial division — or "baby divide," as some practitioners term it — is useful for finding small factors of $N$.

There's trouble, however, if $N$ is something like $2^{193} - 1$. The smallest prime factor is a fairly accessible 13,821,503. The second prime divisor of $N$ lies somewhere in more distant parts. In fact, if a computer could perform a billion division instructions per second, it would require more than 35,000 years of computer time to find the second largest factor of $2^{193} - 1$. It took a new approach and considerable effort by mathematicians Carl Pomerance and Samuel Wagstaff before they could determine that

$$N = 13,821,503 \times 61,654,440,233,248,340,616,559$$
$$\times 14,732,265,321,145,317,331,353,282,383$$

with each factor a prime.

At issue is whether a factoring method can be discovered that works efficiently for any number. Clearly, trial division is good enough for factoring a large number that happens to be the product of many small primes, but so far, a 200-digit composite that is the product of two 100-digit primes seems immune to any known assault.

Two conflicting sentiments lie at the core of factoring theory and practice. On the one hand, mathematicians would like to find new, faster algorithms that would let them factor even larger numbers within a reasonable amount of time. On the other hand, cryptographers who have based the security of their systems on the presumed difficulty of factoring want to be assured that really fast algorithms for factoring large, hard numbers do not exist.

While conflict between these interests continues unresolved, factoring gets steadily speedier. One important advance occurred in 1971, when John Brillhart and Mike Morrison tried a roundabout rather than direct attack on factoring a hard number. They looked for integers that could act as stand-ins for $N$, the number to be factored. Their hope was that these substitutes would be easier to handle yet would provide the necessary information to flush out the primes of $N$.

The idea is to seek two numbers, $x$ and $y$, that satisfy the equation $x^2 - y^2 = N$. In other words, a mathematician hoping to crack a hard composite must find two squares whose difference is the number to be factored. A trivial example illustrates the method. If $N$ happens to be 24, then choosing $x = 7$ and $y = 5$ meets the equation's requirements. Because $x^2 - y^2$ can be factored algebraically into $(x + y)(x - y)$, then $N$ must have the factors 12 and 2. Although 12 can be factored further, this method provides a good start on factoring the number 24. The hard part is finding $x$ and $y$, because random dipping into the pool of available integers is about as futile as locating two needles in a haystack.

What about a situation that requires locating thousands of needles in a haystack that happens to contain millions? That sounds a little easier, especially if the extra work provides a greater chance of success. In fact, mathematicians have created just such a numerical haystack by breaking down the problem of factoring one large number into the problem of factoring thousands of smaller numbers that are relatively easy to manipulate. Because there are millions of such numbers to choose from in any particular factoring problem, the best strategy is to avoid wasting time on any small numbers that prove to be uncooperative. A recalcitrant choice simply loses its place, and another number steps in. That's the idea behind the quadratic-sieve and continued-fraction methods, two of the most important general factoring methods now in use.

The speed of factoring depends not only on the algorithm chosen, but also on the type of computer used. Like the numerous ways in which partitions may divide a building into hallways and rooms to channel movement from place to place, a computer's architecture, built into its circuits and wiring, controls the flow of data. Different computers shuffle digits in different ways. Just because a path can be found from one place to another, whether within a building or a circuit, doesn't mean that it's the most efficient way to go.

Although buildings have movable interior walls and computers are programmable, restrictions remain and bad fits may occur. For example, shoehorning a factoring method such as the quadratic sieve into a given computer is sometimes akin to asking a heart specialist to diagnose a foot ailment. It may take longer than necessary for the answer to appear.

One solution to the problem of computers set in their ways is to select a computer that seems to fit the chosen algorithm, then tinker with the algorithm until it runs efficiently on that particular machine. For example, in the computer version of the quadratic-sieve algorithm are extremely long strings of digits. Such a string, also known as a vector, may have several thousand components. During factoring, this vector is modified thousands of times, but only a handful of digits change each time. Normally, no matter how few digits are changed, the computer's processing time depends on the length of the entire vector. A Cray supercomputer is designed, however, so that changes can be made in specific vector components, and the time taken depends only on the number of changes made, not on the vector's overall length. That feature alone could improve quadratic sieving enormously.

In 1983, Gus Simmons, Jim Davis, and Diane Holdridge of the Sandia National Laboratories implemented the quadratic-sieve algorithm on a Cray supercomputer. Within months, their program was able to take on a 58-digit "most wanted" number and vanquish it w...n 9 hours. Davis then worked out some modifications to the algorithm that further quickened the factoring program, and soon after, one "most wanted" number after another bit the dust.

Marv Wunderlich was able to speed up the continued-fraction factoring method by tailoring it for a special computer called the Massively Parallel Processor, an experimental computer at the NASA Goddard Space Flight Center. This machine has 16,384 simple processors working like a sharply coordinated drill team, so that each performs precisely the same operation at the same time on different bits of data. Wunderlich and Hugh Williams modified the continued-fraction algorithm to take advantage of the parallel processing available, especially for doing the huge number of trial divisions the method requires. They also introduced features that allowed the program to cut its losses early whenever it marched itself into an unprofitable digital corner. The careful fitting of the continued-fraction algorithm to the characteristics of the MPP added 10 digits to the size of numbers that could be factored in less than about 10 hours, bringing it closer in performance to the quadratic-sieving operation on a supercomputer.

Curiously, the five or six best general-purpose factoring methods available seem to share at least one feature. As a result of refinements, all of them seem to have approximately the same theoretical upper limit on their running times. Although such limits are not mathematically rigorous, they represent reasonable estimates of how long a particular algorithm would take to do its job. Whether that limit represents the true level of difficulty for factoring integers isn't clear. No one seems to know what to make of this convergence.

The slow but steady progress in factoring hasn't been nearly as spectacular as the advances in testing for primality. This may mean that factoring really is a difficult problem for which no truly efficient algorithm can exist. On the other hand, a brand new idea could jump the apparent barrier that current methods seem to face. Mathematicians still seem to be a long way from knowing how difficult factoring actually is.

## UNPACKING A KNAPSACK

Secret messages are central figures in a sophisticated, mathematical game of hide and seek. The players are a small group of mathematicians and computer scientists adept at inventing and solving puzzles. They gleefully root out the fatal flaws that may lie hidden within rival encryption schemes while conjuring up new methods that they hope will resist such determined attacks. Their digital games continually stretch the limits of number theory.

To these puzzlers, deciphering a cryptogram such as JXUCQJX-UCQJYSQBJEKHYIJ is a trivial pursuit. In this case, each letter of the original message happens to be replaced by another letter that is a fixed number of places away; for example, an A by a C, a B by a D, and so on. Julius Caesar used this kind of simple cipher more than 2,000 years ago to hide military information. Modern encryption schemes are much more elaborate and mathematically complex than Caesar's simple cipher. But are they unbreakable? The answer to that question is a major concern for the people entrusted with protecting sensitive data from eavesdroppers, thieves, and spies.

Many cryptographic schemes used today seem to be so secure that the only obvious way to break them is by trying every possible cryptographic key until the solution surfaces. An exhaustive search that requires a computer to work out these trial keys for deciphering the message could easily take years, even with the fastest computers available. Alternatively, a computer could consult a ready-made table listing every possible key for a given cipher, but such a table would probably take up an enormous amount of computer memory. It's no simple matter to discover techniques for breaking these schemes.

However, conventional cryptography has a serious flaw that opens a potential window of vulnerability. The sender of secret information must have a key for encrypting messages and the receiver a key for decrypting those messages. The sender must have a secure way of transmitting not only the secret message but also the key needed to unlock it. In other words, any participants in a clandestine operation must all safely exchange decryption keys before they can send and receive messages.

In 1976, Martin Hellman and Whitfield Diffie proposed the notion of public-key cryptography to circumvent the key-exchange problem. In this system, the encryption key is public and available to all senders, while the decryption key is kept secret. The security of this system rests on finding a mathematical way of generating two related keys such that knowing just one of the keys and the encryption method is not enough to recover the second key.

Any mathematical operation that enables such a scheme to work must act like a trap that's much easier to fall into than to escape. Modular arithmetic, in which only leftovers count, provides such a one-way trapdoor. For example, it's easy to compute that 23(mod 5) gives the same remainder as 3(mod 5). However, if the remainder, after division of some unknown number by 5, is given as 3, there's no way of guessing the original value of that mysterious figure. It could be 3, 8, 23, or even 63.

Several public-key cryptosystems have been proposed. One is the RSA scheme, named for its inventors Ron Rivest, Adi Shamir, and Len Adleman. Their method is based on the observation that multiplying together two prime numbers is simple, whereas figuring out the prime factors of a composite number is very difficult.

Using the RSA technique, an individual keen to receive messages would announce a public key consisting of a large number $N$ and an integer $r$. Anyone wishing to send a message to this individual would transform the message into an integer of length $N$, dividing the message into blocks if necessary. The remainder after each block is raised to the power $r$ and divided by $N$ would be sent as the secret message.

The recipient's key is another integer, $s$, to which no one else is privy. Raising the encrypted message to the power $s$ automatically unscrambles the message. The recipient is safe from illicit eavesdropping because the only possible way to compute $s$ requires knowing not just $N$ but the prime factors of $N$ as well. Therefore, the recipient has a way of decrypting a message, but everyone else must first factor $N$ before they can get a crack at breaking any transmissions. If $N$ is large enough, that task is practically hopeless. In contrast, because primality testing is quick, the recipient can readily come up with a suitable 200-digit number $N$ by finding two 100-digit primes and multiplying them together.

An example illustrates the RSA scheme's fine points. An official at the Central Security Agency picks any two prime numbers, say, 3 and 11, and multiplies them together to get 33, the value of $N$. Next, the official goes through a special procedure to come up with the two powers, $r$ and $s$. The procedure's first step is to subtract 1 from each prime to get the numbers 2 and 10, which are multiplied together. The public number, $r$, is any number between 1 and 20 that is not a factor of 20, which eliminates 2, 4, 5, and 10 from consideration. The number 7 is a suitable choice. A headquarters computer finds a number that when multiplied by 7 then divided by 20 leaves a remainder of 1. Modular arithmetic guarantees that there will always be such a number. The number 3 does the job because $7 \times 3 = 21$, and 21(mod 20) equals 1. Hence, $s$ can be 3.

Headquarters is now ready to send out its spies, all equipped with the public numbers 33 and 7. Agent Alice sends a secret message by converting each letter in the alphabet into a number. The number 2, for example, could represent the letter B. To send B, Alice writes down 2, raises it to the seventh power to get 128, then computes 128(mod 33), probably with the help of a handy pocket computer. The secret message is transmitted as 29. A headquarters computer decrypts the message by raising 29 to the third power, then calculating $29^3$(mod 33). The unscrambled message is 2.

The RSA scheme is somewhat cumbersome, and building an efficient integrated-circuit chip to perform the necessary calculations efficiently is tricky. The slow speed at which the RSA cryptosystem operates means a significant bottleneck in transmitting large volumes of confidential data. As a result, mathematicians have also proposed alternative public-key cryptosystems. One that offers faster encryption and decryption is the *knapsack* scheme, first suggested by Hellman and Ralph Merkle.

Knapsack cryptosystems are based on a puzzle known as the knapsack problem: given the total weight of a knapsack and its contents and the weights of the individual items that may be in the knapsack, determine which items are likely to be packed inside so that the total weight adds up to the given amount. Mathematically, the more general problem involves deciding whether some members of a particular collection of positive integers add up to another given integer. If the collection of numbers contains 1, 2, 4, 8, 16, and 32 and the given total is 37, the answer is "yes" because $1 + 4 + 32 = 37$.

In a knapsack public-key cryptosystem, the public key is an ordered set of $n$ "knapsack weights." To encrypt a message consisting of a sequence of 0s and 1s (for example, data stored in a computer), the message is broken into blocks of $n$ bits. Each bit in a block is multiplied by each corresponding number in the public key, and then all these products are added together. The answer is the encrypted message. The method's success or failure depends on the proper selection of these weights in the knapsack.

Merkle and Hellman's idea was to take an "easy" knapsack problem for which a fast method of solution was known and to disguise it by running it through a trapdoor to produce a knapsack that masquerades as a "hard" knapsack, one that takes an incredibly long time to solve. One simple example of an easy knapsack is the special set of numbers 1, 2, 4, 8, 16, and 32. Each number is 1 larger than the sum of all the previous numbers. If the encrypted message is 37, it isn't difficult to discover that the actual message is 101001, because the first, third, and sixth numbers in the knapsack, or public key, add up to 37.

Merkle and Hellman used a generalization of this example, called a superincreasing knapsack, as their set of weights. As a trapdoor, they used a modular multiplication pair. For the pair (28, 71), for instance, each original weight is multiplied by 28 and then divided by 71. Only the remainders are written down. This turns the key into the numbers 11, 41, 22, 56, 28, and 44. To disguise the knapsack further, the order of the items can also be rearranged. Decryption involves finding a second modular multiplication pair (in this case, 33 and 71) that converts the public key back into its original form; the message is again easy to solve. There happens to be a fast way of finding this reverse modular multiplication pair. More complicated schemes — iterated knapsacks — require performing several modular multiplications.

Once the knapsack scheme was proposed, it became the subject of intense scrutiny, and the game began. At that stage, searching for an effective attack on the knapsack cryptosystem was like wandering around in a small piece of a maze. A godlike figure looking down from above can see whether an exit exists, but the lost soul randomly searching within has no idea whether an escape route can be found. Nevertheless, early on, some cryptosystem experts suspected that there might be a way to rip open the knapsack. Nobody knew of one, but they believed that somewhere in the patterns of digits in the encrypted messages and public keys were subtle clues that would make it possible to decrypt any message.

In 1982, Shamir made the first successful attack on the simplest form of the knapsack cryptosystem. He found that certain information about superincreasing sequences is not well disguised by a modular multiplication trapdoor. In addition, this information could be recovered rapidly by solving a special kind of mathematics problem. Soon after, using a similar approach, Adleman broke another form of the knapsack cryptosystem known as the Graham-Shamir knapsack.

Meanwhile, Shamir collected $100 from Merkle, who had offered that sum as a prize for anyone who could break his basic scheme. But Merkle, reacting to the publicity surrounding Shamir's feat and the as-yet incorrect assertion that all knapsack cryptosystems were no longer secure, offered a new $1,000 prize for breaking the iterated knapsack. That was finally done in 1984 by Ernie Brickell, then at the Sandia National Laboratories. Brickell's technique depends on the fact that modular multiplication is the only method being used to hide the knapsack.

This result doesn't rule out the possibility that a secure knapsack cryptosystem can be found. However, it means that a method other than modular multiplication must be used to hide the message. Of course, cryptologists can't resist the challenge of coming up with a cryptosystem that circumvents the flaws pinpointed by Brickell's detection technique. Within months of Brickell's discovery, other schemes appeared, including one based on arithmetic in mathematical structures called finite fields. But the history of cryptography is essentially a history of failures. Lots of cryptosystems have been invented and used only to be proved insecure, sometimes with disastrous results.

Only two serious public-key cryptosystems have been proposed.

The knapsack scheme has already been broken, and the RSA method depends on the complexity of factoring, which remains an open question. Surprisingly, no other fundamentally different techniques have been put forward. Left with only the worrisome RSA scheme, cryptologists and security analysts are frantically seeking new mathematical bases for secure cryptosystems.

One central difficulty in all this activity is that no one can yet prove mathematically that a cryptosystem is secure. All that anyone can say is that a lot of people have poked and prodded a given scheme for several years and that nobody has figured out a way to attack it. Current mathematical techniques are good enough to put a fence around a problem to show where security lies. They can show that the only way in is through a particular gate. But how strong is the lock on the gate? Mathematicians still don't have good techniques for answering that question.

# 3

## TWISTS
## OF
## SPACE

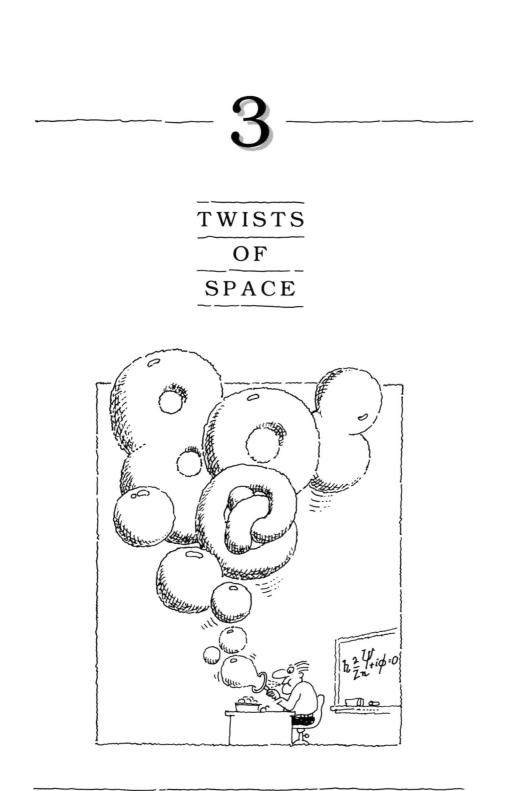

In his charming book on soap bubbles, British physicist Charles Vernon Boys writes: "There is more in a common bubble than those who have only played with them generally imagine." Recent developments in mathematics show that those words, written at the turn of the century, are still true. Inspired by the geometry of soap films stretched across twisted wire frames and soap bubbles clustering in unruly masses, a few mathematicians have ventured into a new world of exotic geometrical shapes.

"Soap films provide a wonderful and accessible physical experiment that leads to many complicated mathematical problems," mathematician Anthony Tromba once said in an interview for a University of California at Santa Cruz publication. "The problem is there. It wasn't invented. Since it does exist, it stands as a challenge to the ingenuity and creativity of the mathematician."

## ── FAIRY-TALE TENTS ──

The structures of architect Frei Otto are contoured as gracefully as soap films stretched over spider webs. Translucent membranes, supported by steel wire nets, reach out from tall masts. Anchors tie the fringes of these ethereal forms to the ground (see Figure 3.1). Their tentlike look is no accident. Otto wanted to use the least possible amount of construction material to create lightweight structures that are easily erected, dismantled, and moved. His models were drawn from nature, in the elegance and economy displayed by soap films. In fact, one of Otto's main tools for designing his exhibition halls, arenas, and stadiums was experimentation with soap films.

The experimenter starts with a plexiglass plate studded with thin rods of various heights. Drooping threads hanging loosely from post to post define rudimentary edges and ridges. Dipping a configuration like this into a soap solution and gingerly withdrawing the contraption magically transforms it into a glistening tentlike shape. The soap film, stretching out only as far as it must, pulls the threads taut to create a spectacular, scalloped roof. Like the contours in Figure 3.1, every section of the roof is shaped like a horse's saddle, curving up in one direction and down in the direction at right angles to the first.

The soap-film models are then carefully photographed and measured. Solid miniatures are built and tested in wind tunnels to determine the potential impact of snow and wind. Finally, after plans are drawn up, construction begins, with sheets of synthetic material re-

**FIGURE 3.1** Architect Frei Otto used soap-film experiments to design the roofs of several Olympic buildings in Munich. *Top,* exterior view of swimming arena; *bottom,* drawing of stadium roof.

placing the soap film and steel cables replacing the threads. With such models, Otto and his collaborators can study forms that have never before been used for buildings. They can explore shapes that are often too complicated to describe mathematically in a precise way. They can solve by experiment numerous mathematical problems associated with surfaces and contours.

Many generations of mathematicians have likewise felt the lure of soap films and soap bubbles. These natural forms embody methods for detecting a surface's optimal shape for a given contour or boundary. Soap films, in particular, are good models for minimal surfaces. Gently poking the surface of a film stretched across a wire loop always increases the film's area. When the disturbance is gone, the soap film springs back to its orginal shape, again taking on the smallest possible area that spans the loop.

The basic physical principle underlying this behavior is a system's tendency to seek a state of lowest energy. Stretching a soap film increases its potential energy. Work must be done to deform it, in the same way that energy is needed to extend a spring. Whenever it can, a soap film seeks a shape that minimizes its potential (surface) energy, and because its surface energy is proportional to its area, it automatically assumes the form of a minimal surface. Consequently, soap films can be used to solve mathematical problems like those that come up in Frei Otto's design work.

Soap-film behavior also hints at ways to solve obstacle problems, which figure prominently in road builders' nightmares — finding paths of least total length, or lowest cost, to link cities separated by turbulent landscapes. A simple version of an obstacle problem has no obstacles and only four cities situated on an unnaturally smooth plain. Devised nearly two centuries ago by Jacob Steiner of the University of Berlin, the problem of finding the shortest route connecting all four cities can be solved using elementary geometry. It can also be solved by building a plastic model consisting of four upright pegs, representing the cities, sandwiched between two transparent parallel plates. When this model is dipped into and then pulled out of a soap solution, the answer is written in the soap film connecting the pegs (see Figure 3.2).

The situation gets more complicated with the introduction of obstacles such as lakes or mountains and by expanding the number of cities. In general, these obstacle problems come up whenever engineers, planners, or managers must find the best path through a thicket of constraints. Whether derived by experiment, worked out on a computer, or extracted using calculus techniques, soap-film solutions show the way.

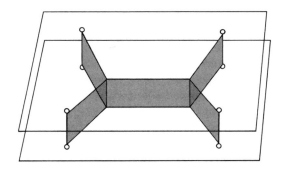

**FIGURE 3.2** The soap-film solution to Steiner's four-cities problem links four points so that the path joining them has the minimum possible length.

## —————— GETTING TO THE SURFACE —— ——

A ring dipped into a basin of soapy water comes out spanned by a shimmering, iridescent film. Of all possible surfaces within this circle, a soap film takes on the form of a thin, flat disk. Any other ring-bounded surface, whether barely wrinkled or strongly bulged, would have a larger area. Thus, a flat disk is the most miserly surface that a soap film spanning a circle can take on without showing some holes. Such a disk is the minimal surface defined by a circle.

When the ring is replaced by a twisted but still closed wire loop, the shape of the corresponding minimal surface is much less obvious. More than a century ago, Belgian physicist Joseph A. F. Plateau spent years experimenting with liquid films and thinking about the forms that emerge. He observed that any loop of wire, no matter how bent, bounds at least one soap film.

Replacing the wire by a curve, or contour, and the soap film by a surface turns a specific physical observation into a broad mathematical question: Does at least one minimal surface—the mathematical version of a soap film—span every conceivable closed curve in space? That general question has become known as the Plateau problem, and it has perplexed mathematicians for decades.

Soap-film experiments show how convoluted this question can get. A loop of wire bent so that it looks like the outline of a pair of old-fashioned, bulky earphones, for example, can come out displaying a soap film with any one of three different configurations (*see Figure*

*3.3)*. Dipped twice into the same soapy liquid, a given wire frame may show two very different forms. It is even possible that a wire frame twisted in the right way would emerge with a different soap-film form every time it is dipped.

Although demonstrations with soap films and twisted wires are fun and give important clues about the possible behavior of minimal surfaces, we can't settle this mathematical question by experiment. The trouble is that no set of experiments, no matter how extensive, can rule out the possibility that doing the experiment one more time would not lead to some new, unexpected result. A mathematical proof, expressed in mathematical terms, is necessary. And it can only be done using mathematical tools.

The study of minimal surfaces lies at the intersection of geometry and analysis, where curve and shape meet tangent and area. The mathematical discipline known as analysis includes calculus and differential equations, which can be used to characterize minimal surfaces. These equations define how a surface changes from place to place, providing the coordinates for a kind of topographic map of the surface in space.

Given the diversity of forms that a soap film or minimal surface can take on, mathematicians have also developed schemes for putting surfaces into different categories, depending on their various features. One rough but useful classification is based on topological type. This comes out of a branch of geometry called *topology*, often described as rubber-sheet or plasticene geometry. In the strange world of topology, where distances have little meaning, a single-handled mug and a doughnut are indistinguishable.

Two geometrical forms are of the same topological type if one shape can be stretched, squeezed or twisted until it looks just like the other. Cutting and pasting or tearing are not allowed. This means

**F I G U R E   3.3**    This example shows that a closed curve can be spanned by three different minimal surfaces.

**FIGURE  3.4**    A coffee mug turns into a doughnut.

that every point in one object finds a place in the other. A line and a circle, for instance, aren't topologically the same because a circle must be torn before points on the circle can be mapped onto a line. A doughnut and a coffee cup are topologically equivalent because it's possible to imagine expanding the coffee cup's handle while shrinking its cup until all that's left is a ring *(see Figure 3.4)*. On the other hand, there's no smooth way to transform an ordinary juice glass into a doughnut without punching a hole in the glass.

By this reasoning, the surfaces of a sphere, a bowl, and a coin all belong to the same topological class or genus, in this case genus 0. A sphere with an attached handle, a doughnut (or torus), and a coffee mug belong to genus 1. A sphere with two handles, a pretzel, and many soup tureens belong to genus 2 *(see Figure 3.5)*. Counting the number of holes in a surface determines its genus. The basic model for each genus can be thought of as a hollow sphere (because only surfaces count) with the appropriate number of handles.

A topological surface, like a sheet of perforated plastic, may also

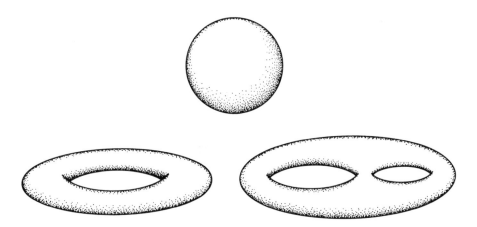

**FIGURE  3.5**    Surfaces of genus 0 (sphere), genus 1 (doughnut or torus), and genus 2 (pretzel).

be punctured in a certain number of places. Such a puncture point is called an *end*. A sphere with one puncture point can be stretched out to form a disk or even an infinite plane *(see Figure 3.6, top)*. Anyone who has widened a clay pot's mouth by pulling on its rim has a sense of how such a mathematical deformation works. In the same way, a sphere with two ends can be transformed into the graceful, infinitely extended hourglass form of a catenoid *(see Figure 3.6, bottom)*. Topological type, then, depends on a surface's genus, or number of handles, and on the number of ends that puncture its skin.

The disk — a sphere with one end — is the simplest possible surface bounded by a closed curve that doesn't intersect itself. The curve itself can be as plain as a circle or as complex as a madly twisted wire, so long as no two of the wire's points touch each other as they would in a figure 8. Such disk-type minimal surfaces are among the simplest solutions to Plateau's problem.

Over the years, mathematicians have been able to prove that every nonintersecting, closed curve is spanned by at least one smooth, area-minimizing surface. Nevertheless, the collection of all possible minimal surfaces turns out to be extremely rich and remarkably complicated, and many are not yet understood. Mathematicians continue to struggle to describe and characterize these surfaces. Classifications of topological type help, but so do measures of curvature, which play a central role in differential geometry.

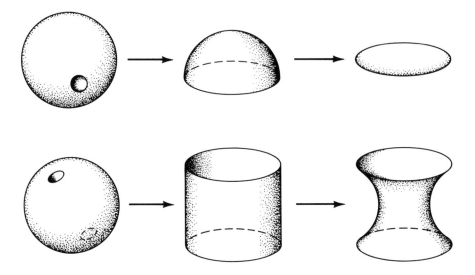

**FIGURE 3.6** *Top*, a sphere with one puncture stretched into a disk; *bottom*, a sphere with two punctures stretched into a catenoid.

Topology focuses on form, whereas differential geometry concentrates on measurable properties such as surface area and the distance between points. Using the ideas of differential geometry, two-dimensional creatures living on a surface could work out the geometry of their environment not by seeing its shape from the outside but by using instruments to measure its local features.

An intelligent ant, for instance, could measure the curvature of its world by pacing off a circle. It would start by defining a reference point. From that starting point, the ant would step out a certain distance in every direction, marking each endpoint so as to draw a circle. The ant would then measure the circle's circumference and compare it with the result of multiplying the circle's diameter by $\pi$.

If the ant's measurement equals the calculated answer, then the surface, by this procedure or test, has a zero curvature. A difference between the measured and calculated circumferences indicates that the surface is curved. The size of the difference, known as the Gaussian curvature, provides an estimate of how curved the surface is. By this measure, a flat sheet of paper has zero curvature; so does the surface of a cylinder, which we can form from a flat piece of paper simply by gluing two edges together. A ball and the inside surface of an ordinary bowl have positive curvature, whereas the surface of a cooling tower and of a holly leaf typically have negative curvature.

A surface of constant curvature is one for which the ant's experiment gives the same result anywhere on the surface. Although both a ball and an egg have positive curvature, only the ball's curvature is constant by this test. It's also possible mathematically to define an average or mean curvature for a given surface.

The concept of curvature helps to characterize minimal surfaces. Because each point on a minimal surface must have a mean curvature of zero, such a surface must be either flat—as on a plane or a cylinder—or shaped like a saddle. In saddles, the surface at each point smoothly curves both away from and toward the point *(see Figure 3.7)*. A four-legged creature standing anywhere on a true minimal surface would find two of its legs sliding down and away from its body and two edging down and closer to its body. That would happen at any spot on a minimal surface, unless the surface happened to be perfectly flat. In terms of soap films, saying that the mean curvature must be zero is equivalent to noting that the pressure on both sides of a soap film is the same.

Another important measure is that of total curvature. An ant following a circular path turns itself through 360° over the course of one complete circuit. That total amount of turning can be expressed as the number $2\pi$, which represents the total curvature of any circle, regardless of its radius.

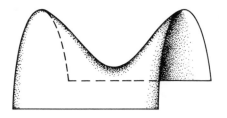

**FIGURE** 3.7    A saddle-shaped surface has a mean curvature of zero because the surface curves both downward and upward in the vicinity of every point.

Twisted, closed loops of wire have a total curvature that may be greater than the curvature of a simple circle. A wire that loops around twice before its ends join, for instance, has a total curvature of $4\pi$. As one step in solving the Plateau problem, mathematicians have been able to prove that for curves or contours with a total curvature of less than $4\pi$, only one disk-type minimal surface is bounded by that contour. However, if the total curvature of a curve is even slightly larger than $4\pi$, then quite wild and unimaginable things can occur.

Surfaces also have a total curvature. In the case of a sphere, it turns out to be $4\pi$ and is again independent of the radius. One method for computing the total curvature for any surface applies a remarkable theorem that connects differential geometry and topology. Established before the turn of the century by Bernhard Riemann and David Hilbert, the theorem states that any surface is topologically equivalent to a surface of constant curvature. Hence, a rubbery egg and even a flexible cube can always be deformed into a perfectly round sphere. Both objects have a sphere's total curvature—the cube's curvature is simply concentrated in its edges and corners. No matter how much the object is deformed, its total curvature depends only on its topological type.

The surface of a doughnut—topologically a torus or a sphere with one handle—turns out to have just as much positive curvature as negative curvature. A torus is just a cylinder that's been bent so that its two ends meet, just as a tube of dough can be used to make a bagel. This figure's total curvature is zero. Adding a handle to a torus to make a pretzel decreases the surface's total curvature to $-4\pi$. In fact, every additional handle makes the surface's total curvature more negative by $4\pi$.

With the concepts of topological type and curvature, mathematicians can begin to explore minimal surfaces and search for patterns that could lead to theorems about minimal surfaces. Still, progress is slow. Many questions remain unanswered. No one yet knows how to estimate the number of different minimal surfaces that may exist for a given closed curve. Even the much simpler question of how many disk-type minimal surfaces can span a closed curve isn't completely settled.

Meanwhile, topologists have lots of other intriguing surfaces to play with and ponder. One such minimal surface, discovered in 1977 by Bill Meeks during a bout of mathematics at the Gaslight Coffee Shop in Amherst, Massachusetts, contains a Möbius strip. Constructed by gluing together the two ends of a long strip of paper after giving one end a half twist, a Möbius strip has only one side and one edge *(see Figure 3.8)*. The minimal surface containing it twists in space, cutting through itself and reaching out to infinity in all directions *(see Color Plate 1)*. That strange surface would be enough to dizzy any ant explorer.

————————— BITES IN A DOUGHNUT —————————

The disklike soap film clinging to a ring emerging from soapy water is just one small piece of an idealized, mathematical object called the plane. A plane's vast surface extends over an infinite area. Yet at the

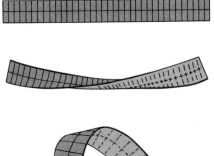

**FIGURE 3.8** A Möbius strip can be constructed by joining the two ends of a long, narrow strip of paper after giving the strip a half twist.

same time, it can still be considered a surface of least area. In some ways, it's like a giant soap film whose boundary extends so far over the horizon that the boundary can't be seen. We must find the evidence for its economy in local clues.

Like our ant in the previous section, any creature striding along an infinite plane can check the minimal nature of its surroundings simply by outlining a closed curve anywhere on this surface. That piece of surface will have the smallest possible area bounded by the curve. If the test works everywhere, then the creature can be reassured that it is, indeed, wandering across an infinite minimal surface.

Two rings dipped in a soap solution may each come out with a disklike film. But a soap film connecting the two rings may also emerge, creating a minimal surface that looks like a pinched cylinder, smoothly taken it at its waist. That hourglass form is the central piece of another infinite minimal surface called a catenoid *(see Figure 3.9, left)*. Its two open mouths, one at either end, reach out infinitely far into space.

Mathematically, a catenoid can be defined in terms of the curve formed by a hanging chain fixed at both ends. Rotating this dangling curve, called a catenary, about a horizontal straight line produces the central section of a catenoid lying on its side.

A loose coil of wire twisted into the form of a helix supports a spiraling soap film. Extending its ends creates an infinite slide that

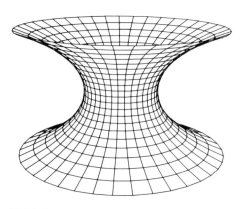
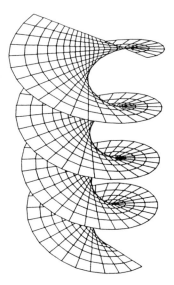

**FIGURE 3.9** *Left,* catenoid; *right,* helicoid.

would whirl any plummeting rider along its surface into a dizzy spin. This infinite minimal surface is called a helicoid *(see Figure 3.9, right)*.

The plane, catenoid, and helicoid share two additional features. Because they are, roughly speaking, boundaryless, they are called *complete* minimal surfaces. Because none of them folds back and intersects itself, all three are said to be *embedded* minimal surfaces. Until recently, the plane, catenoid and helicoid were the only known examples of complete, embedded minimal surfaces of finite topology. Topologically, both the plane and helicoid can be modeled on a hollow sphere with no handles and one hole, whereas the catenoid resembles a twice-punctured sphere. A few mathematicians had even speculated that these were the only examples that could meet such stringent mathematical specifications.

Topologists had other reasons to suspect that perhaps more than three such surfaces actually existed. The trouble was that potential candidates for this minimal-surface hall of fame typically were expressed in complicated equations that masked more than they revealed. This inside information had to be unlocked before the surface's true nature could be seen. Particularly vexing was the problem of determining from the equations whether a surface is embedded, or folds back on itself. Moreover, because these elusive minimal surfaces are infinite in extent, soap-film experiments would be of little use.

The story of the discovery of whole families of new minimal surfaces of this special type begins with a set of equations first written down by Celso Costa, a Brazilian graduate student. Costa had managed to prove that his thorny equations represent an infinite minimal surface. Topologically, Costa's surface can be modeled on a three-ended hollow sphere with one handle, or in more tantalizing terms, on a chocolate-covered doughnut from which three bites have been taken. The bites puncture the surface and indicate that the surface, when deformed, can extend to infinity in three places.

But the question remained: Did the surface intersect itself? Mathematicians David Hoffman and Bill Meeks of the University of Massachusetts then took up the quest. They hoped that the surface didn't intersect itself, but proving it was no simple matter. Given a set of mathematical equations, there's no single quantity that can be computed to give a yes-or-no answer to the question of self-intersection. All that can be shown is that a particular piece of the surface doesn't intersect another piece. That's not enough for an infinite surface, where an infinite number of pieces would have to be compared.

Hoffman's plan was to use a computer to find numerical values for the surface coordinates and then draw pictures of its core. These "snapshots" might catch the equations in revealing poses. Mathematical clues already hinted that the surface contains a plane and two catenoids that somehow sprout from the figure's swiss-cheese center, but it was hard to see what was happening in the middle.

For assistance in his search, Hoffman turned to James Hoffman (no relation), a graduate student at the university who happened to be working on a new computer-graphics program and language. Together, they spent many hours tinkering with the equations and the software necessary to bring the surface to life on a video screen. David Hoffman knew that if he saw evidence of the surface intersecting itself, then the surface would not be embedded and this particular mathematical quest would be over. If there was no visible evidence of an intersection, then he could go ahead and try to prove that the surface really was embedded.

The first pictures were full of surprises. The surface appeared to be free of self-intersections, and it seemed to have a high degree of symmetry. It contained two straight lines that met at right angles. "Extended staring," in Hoffman's words, led him to see that the surface consisted of eight identical pieces that fit together to make up the whole figure.

The entranced mathematicians started to explore the surface graphically, rotating it, climbing inside it, examining it segment by segment. Many views later, the true form of this minimal surface began to emerge. And it was strikingly beautiful. The figure had the splendid elegance of a gracefully spinning ballerina flinging out her full skirt so that it whirled parallel to the ground. Gentle waves undulated along the skirt's hem. Two holes pierced the skirt's lower surface and joined to form one catenoid that swept upward. Another pair of holes, set at right angles to the first pair, led from the top of the skirt downward into the second catenoid (see Figure 3.10 and Color Plate 2).

The symmetries revealed in pictures of the figure provided Hoffman and Meeks with just the tips they needed to analyze the equations. That, in turn, led to a mathematical proof that the surface, indeed, was the first complete, embedded minimal surface of a finite topology to be found in nearly 200 years. Its predecessors, the catenoid and helicoid, had been discovered in the eighteenth century.

The search didn't end there. Meeks and Hoffman soon demonstrated the existence of an infinite number of surfaces, each one topologically equivalent to a thrice-punctured sphere with a certain number of handles (see Figure 3.11). Where there were once only three complete, embedded minimal surfaces, now there are too many to count, and the number continues to grow. A four-ended minimal surface that looks like the original one cut off at the bottom and then

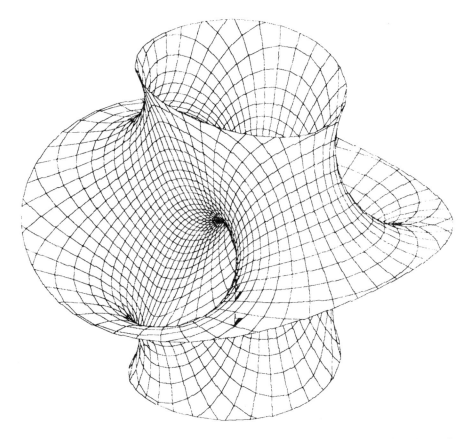

**FIGURE 3.10**  Hoffman's genus-1 minimal surface doesn't intersect itself anywhere. Mathematicians say such a surface is "embedded."

reflected in a mirror has recently been discovered. It starts off a second family of complete, embedded minimal surfaces. In this case, the computer was used not only to draw the images but also to search for the correct values of a parameter crucial to the definition of the surface.

In general, the computer can be used in situations where physical experiments aren't possible. Computer graphics plays the same role for infinite minimal surfaces that soap films play for surfaces spanning a wire contour. It's hard to imagine a wire frame that adequately models surfaces of infinite extent. But computer experiments are themselves limited in at least one important way. An explicit equation defining a surface or group of surfaces is needed to produce a computer picture. Finding such an equation for a surface spanning a

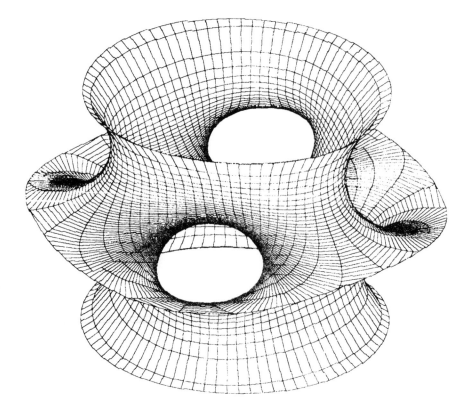

**FIGURE** 3.11    Hoffman's genus-3 minimal surface is topologically modeled on a sphere with three handles and three punctures.

given contour is almost impossible. In contrast, soap-film experiments instantly produce at least one minimal surface for any wire loop. Nevertheless, when an explicit equation is known, the computer is a useful tool for exploring the defined surface. Because the visual exploration often furnishes clues that can be used later for nailing down the mathematical proof, the machine serves as a guide in the construction of a formal proof.

The new family of minimal surfaces may be of more than mathematical or aesthetic interest, for area-minimizing surfaces often occur in physical and biological systems, especially at the boundaries between materials. These newly discovered forms may be useful models for, as one biologist suggests, the shape of developing embryos or for the structure of certain polymers. A dental surgeon has suggested that such a shape could be used in bone implants for securing false teeth. An implant's design with lots of handles and a least-area surface would minimize contact with bone while ensuring a strong bond.

Bit by bit, the strangely beautiful new world of minimal surfaces is emerging. Mathematicians probing complex equations are coming up with fixtures for a surreal plumbing-supply store: contorted tubes, helicoids with tunnels, punctured basins. Some forms blossom into bizarre flowers, while others seem ready to fly off the pages of a science fiction novel. Computer graphics is proving to be an indispensable tool that opens up for exploration a hitherto unseen realm of geometric forms.

— —— ——————— BUBBLING AWAY —— ——— —

At first glance, a basinful of shimmering soap bubbles seems to be nothing more than a haphazard collection of pliant, transparent balls randomly poking into and crowding one another. Nevertheless, a delicate balance holds throughout this ephemeral architecture. The popping of even a single bubble triggers a quick readjustment, then all is still again.

A solitary soap bubble's shape is governed mainly by a force known as surface tension, which is uniform over the whole bubble. In a sense, surface tension represents an elastic skin that preserves the bubble's shape. A soap film enclosing a parcel of air stretches only as far as it must to balance the air pressure inside the bubble — a manifestation of the equilibrium between air pressure and surface tension.

The bubble is spherical because a sphere has the least possible area for the volume that it encloses. Any other shape encompassing the same volume requires a soap film of larger area. Creating that larger area means stretching the film and therefore proportionately increasing its surface energy, in the same way that poking a balloon increases its potential energy. Hence, a soap bubble's perfectly round shape minimizes the bubble's surface energy.

A few rules appear to govern the way soap bubbles cluster together. It turns out that only three things can happen when soap films meet, even in the largest bubble clusters. First, a smooth sheet of film can separate space locally into two regions, as shown when two bubbles are brought together (see Figure 3.12, top left). Second, if three surfaces meet, as they do when three bubbles come together, they intersect to form a line, and the angle between each pair of sheets is 120° (see Figure 3.12, top right). Finally, six surfaces can meet, three at a time, along four edges that come together in a point. That configuration appears when a fourth bubble is placed atop a

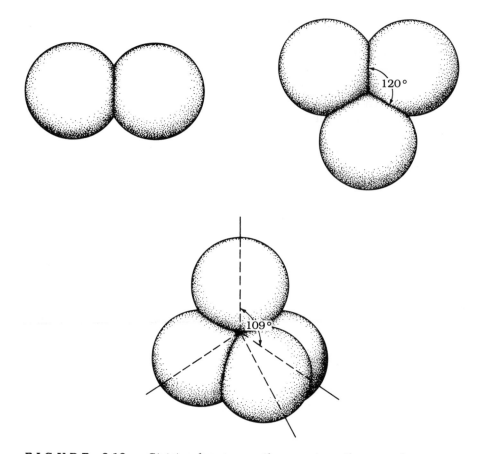

**F I G U R E  3.12**  Strict rules govern the way two, three, or four soap bubbles come together. Two bubbles form a line; three, a triangle; four, a tetrahedron.

triangle of three bubbles so that the whole arrangement looks like a tetrahedron. The angle between pairs of edges is always roughly 109.47°, which corresponds to the angle between two crossing diagonals, each drawn from opposite corners of a cube *(see Figure 3.12, bottom)*.

One of the mathematical achievements of the last decade was the creation of a model for the rules governing the geometry of soap froths and films. The three basic rules of bubble behavior are a mathematical consequence of a simple area-minimizing principle. For soap bubbles and films, that principle is related to a physical system's tendency to seek a minimum surface energy at an interface. Such a system will remain in a certain pattern only if it cannot readily shift to one with less energy.

The change in shape can be seen when two bubbles are brought together. Each bubble keeps its spherical shape until the two touch, at which instant the films flow together, eliminating part of the outer surface of each bubble. Instead of having four separate surfaces, as the bubbles would have had if they had been pressed together without joining, the bubbles now share a single film. This reduction in film area decreases the total surface area and energy of the original configuration. If the bubbles are of equal size, the interface is flat. If one bubble is larger than the other, the boundary film is a smooth curve that bulges toward the larger bubble.

In mathematical terms, according to the model developed by Jean Taylor and Fred Almgren, any geometric arrangement of two-dimensional surfaces is soap-bubblelike if it encloses one or more regions of space and if it cannot be forced by a small deformation into a configuration of smaller area without altering a region's volume. That model has allowed mathematicians to explore a range of questions, from the behavior of soap bubbles to the shape of minimal surfaces that span given curves.

Soap bubbles, in turn, sometimes provide a convenient model for even more complicated physical processes. A metal's etched surface, for instance, typically shows a jumble of grains jammed together. Each grain is a single crystal, made up of atoms in orderly arrays. When the metal solidifies, microscopic crystals that form within the liquid grow until they bump into their neighbors. The subtle interplay between physical and chemical forces and the geometric requirements of filling space sets the final grain boundaries (see Figure 3.13, left).

This grain structure, a natural result of crystal growth, often looks a lot like a soap froth (see Figure 3.13, right) and sometimes behaves like it. Steel, for instance, is a mixture of carbides and iron. When steel is heated up, grain boundaries shift. The grains act much like bubbles clustered together, where larger bubbles grow at the expense of smaller ones to create a coarser pattern. Such observations have led metallurgists to use soap froths as a rough model for a metal's grain structure. The model helps them understand the behavior of metals and suggests ways of manipulating grain structure to get metals with the right properties. This manipulation of microstructure is a central feature of modern materials science.

Yet crystals differ from soap bubbles in several ways. Crystal surfaces lack the flexibility of soap films; they don't readily bend around corners. Instead, these surfaces tend to be flat and to take on definite directions. A crystal's rigidity affects grain boundaries in

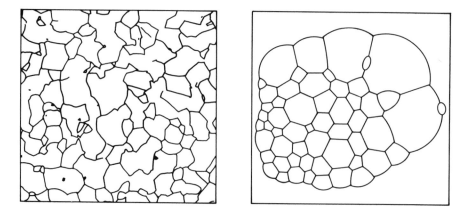

**FIGURE 3.13** *Left,* a deeply etched section of a piece of niobium metal shows a network of grain boundaries; *right,* the froth of irregular soap bubbles shows a cellular structure analogous to that of metals.

ways that the soap-bubble model cannot explain. Taylor has recently been working with metallurgist John Cahn of the National Bureau of Standards to get a better idea of the types of boundaries that can form between adjacent crystals. They are developing a new kind of mathematics for dealing with what, in effect, are cubic or polyhedral bubbles—forms that have well-defined faces. That's just the kind of mathematics that might apply to crystalline grains in metals.

As in the case of soap bubbles, the principle of minimizing surface energy determines the boundaries between crystals and the boundaries between a crystal and a surrounding fluid. For crystals, the surface-energy value depends on the nature of chemical bonds left dangling at particular surfaces. In some directions, the energy required to break apart a crystal may be much lower than in other directions. A crystal's surface energy, then, would be *anisotropic,* which means that it varies from face to face.

Just as a sphere is the equilibrium shape of a single soap bubble, each anisotropic crystal has an equilibrium shape of its own. A crystal's unique equilibrium shape is the one that has the least total surface energy for a given enclosed volume. This special form, the anisotropic analog of a sphere, is often called the Wulff shape, named for crystallographer Georg Wulff, who suggested the idea in 1901.

Whereas soap bubbles and liquid droplets are all spherical (at least in the absence of gravity and other outside influences), crystals with different chemical compositions may have widely varying surface-energy distributions and therefore different Wulff shapes. Depending on the material, a Wulff shape may have the form of a cylinder, a cube, an octahedron, or some other polyhedron. The equilibrium shape of a single crystal surrounded by a fluid or jammed against other crystals may be a polyhedron rather than a sphere.

One of the first steps in studying the border regions between crystals is to classify and catalog what can happen when an anisotropic surface has a given boundary. In other words, what is the crystal analog of a soap film on a wire frame? Plateau's problem of finding minimal surfaces for all possible curves strikes again, but this time in a new, more complicated guise. What are the surfaces of least surface energy spanning a given boundary? Do solutions always exist? Are they unique? How smooth are they? What is the structure of singularities—places where the surfaces are not smooth? How might one compute the minimizing surfaces, given a reasonably smooth, uniform curve as a boundary?

One useful way to reveal these minimal surfaces is to do the mathematical equivalent of letting a spotlight play over a surface to highlight part of an interface. That means fixing a boundary curve that isolates part of the surface in order to determine what types of local structures have minimal surface energies. The result is a complete list or an illustrated catalog of all the interfaces that can occur between a crystal and a surrounding medium, whether solid, liquid, or gas.

Taylor and Cahn worked with a Wulff shape in the form of a cube with lopped-off corners *(see Figure 3.14).* This truncated cube, in

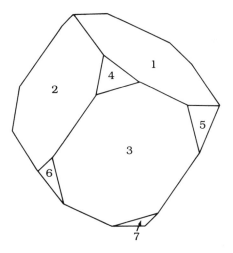

**F I G U R E   3.14**    This particular Wulff shape appears in the form of a truncated cube.

various guises, represents most common crystal forms. For the truncated cube, they found 12 possible types of structures that locally minimize surface energy when an interface separates two regions. In contrast, there's only one possible boundary—a smooth, planar interface—between two neighboring soap bubbles.

Looking like bites taken out of curiously cut diamonds, the 12 types of minimal surfaces corresponding to the truncated cube Wulff shape fit into four main categories (see Figure 3.15). One category consists of forms that look like shapes a spotlight would catch if its light happened to fall entirely on one surface of the Wulff shape (a flat disk), on an edge (two planes meeting), or on a corner (three planes coming together). The other three groups feature various saddle shapes. The effect is not unlike the different ways in which wedges can be cut from a chunk of cheese.

In all 12 cases, the plane faces, usually seen as wedges of the possible minimal surfaces, run parallel to corresponding faces in the Wulff shape (see Figure 3.16). Furthermore, two minimal-surface segments meet along a line only if the Wulff shape's matching faces meet along an edge. Saddle-shaped surfaces have the additional property that if the plane segment on one side of a given wedge bends up, then either the other side bends down or the wedge has a central angle of more than 180°. Whenever a wedge has a central angle of 180° or less, with each neighboring wedge bending down, then the corresponding faces on the Wulff shape must meet at a single corner.

Several of the interfaces shown in the catalog, which Taylor and Cahn produced only a few years ago, are new to metallurgists. The mathematical proof of their minimality corrects a number of misconceptions that have appeared in metallurgical research papers and other writings. The findings suggest that some geometries, which metallurgists believed were caused by crystal defects, are actually forms that arise naturally in the course of crystal growth within a solidifying metal. Furthermore, the catalog answers at least one aspect of the question of why the microstructure of a given material is as it is. Some features of a metal's crystal structure, once thought to be mysterious or the result of some accident, turn out to be completely reasonable structures that obey all the known rules.

In subsequent work, Taylor and Cahn also unexpectedly discovered a cusp-shaped surface that abruptly changes direction at the same time as its surface energy is minimized. One cusp of this type looks like a ledge that peters out as it runs along halfway up a vertical wall (see Figure 3.17, left). Such cusps have actually been seen on crystal surfaces (see Figure 3.17, right), but metallurgists have usually interpreted them as the result of defects or nonequilibrium crystal growth.

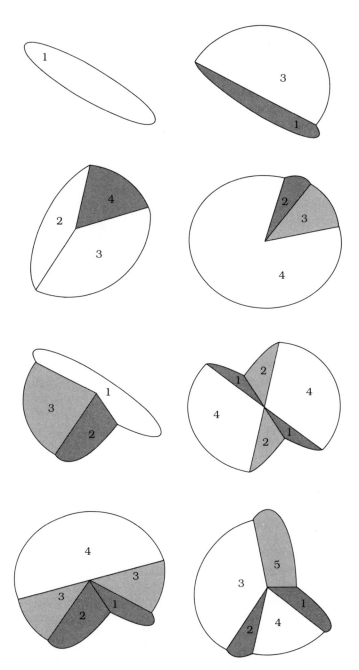

**FIGURE 3.15** Examples from the complete catalog of local structures in equilibrium surfaces of a crystal whose Wulff shape is a truncated cube. Regular edges are shown darker than inverse edges.

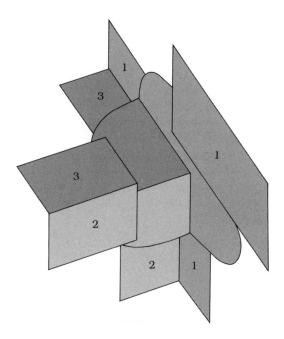

**FIGURE 3.16** The interface structures in the Taylor and Cahn catalog are constructed by considering the possible arrangements of adjacent planar surfaces of the crystal's corresponding Wulff shape. The structure shown here matches the orientation of planar faces 1, 2, and 3 of the Wulff shape.

Taylor and Cahn proved mathematically that a surface of least surface energy can contain a cusp. The presence of a cusp on a crystal surface, they suggest, need not imply that there is a dislocation in the body of the crystal or that the surface is not at equilibrium. The Taylor-Cahn proof implies that cusps can occur in crystals under equilibrium conditions.

Taylor's mathematical contribution goes well beyond applications in metallurgy. Her foray into "cubic" bubbles and related anisotropic forms is an extension of centuries of research, inspired by soap-film studies, on isotropic minimal surfaces. The same questions that can be asked about soap films also apply for anisotropic surfaces. Most of those questions aren't yet answered.

Given a problem as old as Plateau's problem, remarkably little is known about shapes for which the surface energy is assumed to be anisotropic instead of isotropic. Perhaps the reason is that soap films

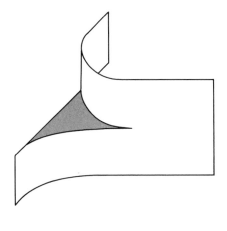 

**FIGURE 3.17** *Left,* a cusp; *right,* a photograph showing cusp forms on the surface of a crystal.

and bubbles have been visible to mathematicians, but the more complex versions of the analogous surfaces for crystalline materials were rarely noticed until relatively recently because the technology to see them was lacking. Furthermore, basic questions concerning the way surfaces change when the surface energy changes — for example, through changing the temperature or surrounding atmosphere — remain unanswered. And what about a crystal sitting on a table? Instead of remaining perfectly regular, it sags ever so slightly under the influence of gravity, distorting its geometric form. The study of anisotropic surfaces is a new field in which much more is unknown than is known.

Other mathematicians are now peering into even more exotic "bubbles" that, unlike spheres, have no counterpart in nature. These strange forms were discovered in 1984 when mathematician Henry Wente of the University of Toledo found that the sphere is not the only finite surface having a uniform curvature.

Until Wente's work, no one knew whether a sphere is the only surface with a constant mean curvature. An egg's surface doesn't fit the definition because its surface curvature changes from place to place. An endless cylinder that stretches to infinity in both directions has a constant mean curvature, but it isn't finite. Indeed, mathematicians had been able to prove that a sphere is the only convex surface of constant mean curvature, but that didn't exclude the possibility of a more convoluted surface fitting the bill. Mathematicians also knew that any sphere with one or more handles can't have a constant mean curvature unless the surface intersects itself. Circumstantial evidence seemed to go against the possibility that such a surface existed. But without a mathematical proof one way or the other, the possibility of another constant-curvature form couldn't be ruled out, no matter how great the prestige of those mathematicians who scoffed at the notion.

Wente's discovery of a strange bubble with constant mean curvature and his proof now settle the question. Modeled on a torus or a sphere with one handle, his three-lobed toroidal soap bubble bends so sharply that the surface passes through itself in an intricate pattern and hides away its core. The only way to see its full geometry is to slice the surface open and peel it away layer by layer, opening it up like an exotic onion (see Color Plate 3). Computer graphics allows a clear view of the inner works of the three-lobed toroidal soap bubble, one of a family of finite nonspherical surfaces of constant mean curvature now known to exist.

## KNOTTY PROBLEMS

In studying knots, mathematicians can get as tangled up in their work as any frustrated individual unraveling snarled fishing line, yarn, or string. In fact, a piece of knotted string with its ends spliced together so that it can't be untied is an excellent physical model for the abstract object with which mathematicians work. A mathematical knot can be thought of as a one-dimensional curve that snakes through three-dimensional space, then catches its own tail to form a loop. Like a circle, such a curve starts and ends at the same point and never intersects itself (see Figure 3.18).

Curiously, "knottedness" is not a property of the curve itself. An imaginary ant crawling along a narrow tunnel within the one-dimensional space of such a curve, even after completing its circuit, would never be able to tell whether the curve is knotted. Knottedness resides in the way the curve sits in three-dimensional space. It must reside in three-dimensional rather than two- or four-dimensional space. In two dimensions, there isn't enough room for the curve to be knotted, whereas in four dimensions there's so much room that any respectable knot can be untied.

Like many aspects of soap-film and soap-bubble geometry, knot

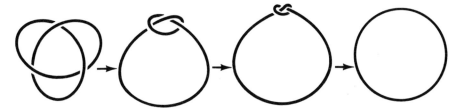

**F I G U R E   3.18**    Models of knots are constructed by thickening one-dimensional curves slightly so that they seem encased in a solid, flexible three-dimensional tube. Otherwise knots can always be squeezed out.

theory is part of the mathematical field of topology, in which smoothness, size and shape can safely be ignored. The only geometric properties that survive are those oblivious to bending, squeezing, twisting, stretching, and other deformations of space. Complete flexibility is the rule. To topologists, different knots, no matter how twisted or tangled, merely represent various ways of embedding a circle in three-dimensional space.

Mathematicians can ask the same questions about a knotted curve that a boy scout may ask about a knotted rope. What kind of knot is it? Is the curve (or rope) really knotted? Can a second knot undo the first? Is one knot equivalent to another? This last question raises the fundamental problem of knot theory: How can different knots be distinguished?

In general, it's hard to tell if a certain knot tied in a string is the same as a seemingly different one tied in another string *(see Figure 3.19)*. One way to solve the puzzle is to try transforming one knot into the other using steps that twist or otherwise deform but don't cut the knot. The trouble with this solution is that it depends on the patience of the person doing the untangling. Spending hours of fruitless labor untwisting and unlooping a knot that still doesn't match its compan-

**FIGURE 3.19** These two 10-crossing knots look different, but some energetic pulling and twisting transforms one into the other.

ion doesn't prove that the two knots are different. The right combination of moves could somehow have been overlooked. It's the same conundrum that faces a person who must decide whether to continue trying to unravel a tangled ball of yarn or to resort to a pair of scissors.

To attack the problem of classifying and distinguishing knots, mathematicians have adopted a set of rules that make knots more convenient to study. Instead of analyzing three-dimensional knots, they examine two-dimensional shadows cast by these knots. Even the most tangled configuration can be shown as a continuous loop whose shadow winds across a flat surface, sometimes crossing over and sometimes crossing under itself. In drawings of knots, tiny breaks in the lines signify underpasses or overpasses, while arrows indicate the direction of travel around a loop *(see Figure 3.20)*. Furthermore, just as a suspended wire frame, caught in a breeze on a sunny day, casts an everchanging shadow on the ground, a knot illuminated from different angles can display various projections on a plane. Mathematicians usually work with the simplest projection they can find for a given knot.

A loop without any twists or crossings—in its simplest form, a circle—is called an *unknot*. Loops that show up as projections with only one or two crossings can always be transformed into unknots *(see Figure 3.21)*.

The simplest possible knot is the overhand or trefoil knot, which is really just a circle that somehow winds through itself. In its plain-

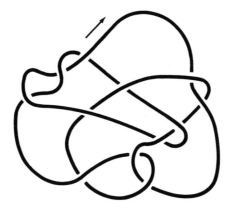

**FIGURE** **3.20**    By breaking the lines in a drawing of the shadow of a knot and by showing a preferred direction, mathematicians can keep track of a knot's spatial placement.

est form, this knot has three crossings. It also comes in two flavors: left-handed and right-handed configurations that are mirror images of each other *(see Figure 3.22)*. No possible deformations of the trefoil knot ever eliminate the crossovers. Only breaking the loop does the job.

There's just one distinct knot with four crossings, and only two with five. The club expands rapidly from there to a total of 12,965 distinct knots with 13 or fewer crossings, excluding mirror images.

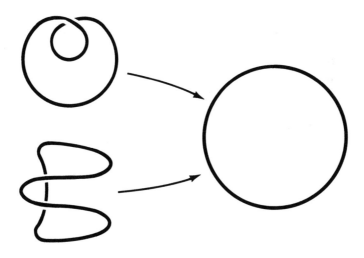

**FIGURE** **3.21**    An apparent knot can be converted into an unknot.

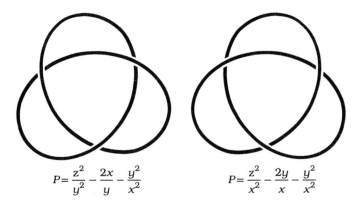

$$P = \frac{z^2}{y^2} - \frac{2x}{y} - \frac{y^2}{x^2} \qquad\qquad P = \frac{z^2}{x^2} - \frac{2y}{x} - \frac{y^2}{x^2}$$

**FIGURE 3.22** The trefoil knot is shown in left-handed and right-handed configurations, with appropriate polynomial labels.

Thirteen is the highest number of crossings for which a complete catalog of knots now exists. Because knots may also be intertwined, like links in a chain, the complexities rapidly multiply.

Sailors, knitters, and other people who regularly work with knots have long had ways to classify knots in terms of their physical characteristics, but the mathematical classification of knots goes back only to the late nineteenth century. British scientist Lord Kelvin hypothesized that atoms were knotted vortices in the ether, an invisible fluid thought to fill all space. By classifying knots, he hoped to organize the known chemical elements into a periodic table. Kelvin's atomic theory died, but the study of knots survived.

The earliest knot theorists compiled huge tables of knots, drawing pictures to show which knots are different from others. But that procedure is cumbersome, tedious, and unsatisfactory. What mathematicians really want and are still looking for is a simple way to pin labels on knots so that two knots with the same label are the same, while two knots with different labels are truly different. In the latter case, the label would be enough to indicate that no amount of twisting, pulling or pushing would ever transform one knot into the other. Easily computed labels would allow knot theorists to tell many knots apart without having to go through the messier task of tangling with the knots themselves.

A mathematical label corresponding to any knot property that remains unchanged by deformations is called an *invariant*. One possible example of such an invariant is the minimum number of crossing points found in a drawing of a knot. This number often serves as the basis for organizing lists of knots. In some ways, however, the crossing-point number is an unsatisfactory invariant. It's not always easy to tell whether a knot has actually been drawn with the minimum number of crossing points. Hence, the number may be difficult to compute. Moreover, this invariant doesn't discriminate very well because many different knots have the same crossing-point number. And a single number holds little information about the structure of a given knot.

Another approach is to use the arrangement of crossings in a knot diagram to produce an algebraic expression that serves as a label for the knot. In 1928, John W. Alexander discovered a systematic procedure for generating such a formula. Expressed as positive or negative powers of some variable with integer coefficients, his simple polynomials for characterizing and labeling knots turn out to be remarkably useful and relatively easy to compute, though not ideal.

If two knots have different Alexander polynomials, then the knots are definitely not equivalent. For instance, the trefoil knot carries the label $t^2 - t + 1$, whereas a figure 8 knot is $t^2 - 3t + 1$. But knots that have the same polynomial aren't necessarily equivalent. The procedure doesn't distinguish, for example, between left-handed and right-handed knots.

It was several decades before mathematicians fully understood the conceptual basis for the calculation of Alexander polynomials and gained a sense of which knot properties the polynomial captures. In Alexander's method, only the direction of the crossings, over or under, and their arrangement with respect to the other crossings make a difference. The formula is the mathematical equivalent of systematically snipping the two strands of the knot at each crossing and refastening the ends so that they are no longer twisted.

In 1963, mathematician John H. Conway, exploiting this new understanding, developed an easier method for computing the Alexander polynomial. Conway's method recognizes that a knot can be progressively unknotted by changing selected over and under crossings. Step by step, his unknotting game leads to diagrams with fewer crossings and finally to the Alexander polynomial. This invariant seemed to be the end of the story, and many topologists interested in knot theory focused their efforts over the next two decades on understanding the Alexander polynomial better and using it more efficiently. In 1984, however, knot theorists were suddenly and unexpectedly thrust into new mathematical territory overrun with novel invariants.

The mathematician who triggered the stampede was Vaughan F. R. Jones of the University of California at Berkeley. He found a completely new invariant, another polynomial that does a better job than the Alexander polynomial. Unlike the method for finding an Alexander polynomial, the Jones approach was based on the idea that overpasses and underpasses (or positive and negative crossings) play different roles. His discovery prompted a great deal of excitement in the mathematical community because his polynomial detects the difference between a knot and its mirror image. Jones also caught the mathematics community by surprise because he specializes in an area that has little to do with knot theory. He unexpectedly linked the abstract world of von Neumann (or operator) algebras, which originated in quantum mechanics, and the low-dimensional topology of knot theory.

News of the Jones polynomial set off a wave of mathematical activity. It seemed that just about every mathematician who saw the letter that Jones circulated announcing his discovery and understood its significance caught a glimmer of new possibilities. Mathematicians raced off to look for a general expression that encompasses both the Jones and Alexander polynomials.

The fact that someone reported success in discovering a way to generalize the Jones discovery turned out to be much less of a surprise than the fact that five independent groups of mathematicians arrived at essentially the same result in practically a dead heat. They all found new polynomial invariants that were even more successful than the Jones polynomial in distinguishing between distinct knots. In some mysterious way, these new, improved invariants seemed to delve deeper into the essence of knots.

It was evident that the mathematicians submitting papers had arrived at their results completely independently of each other, although all were inspired by the work of Jones. The whole situation had the potential for a major dispute over priority, but in the end, it was resolved reasonably amicably. The competing groups eventually agreed that it would be unproductive to try to assess priority. Their answer was to publish one joint paper with six listed authors. A mathematician not directly involved wrote an introduction, and four groups presented summaries of their proofs. The fifth group, a pair of Polish mathematicians, was the victim of slow mail and missed being included in the joint paper.

There are many instances of researchers making the same discovery at the same time, but the combination of so many different people with such a dramatic discovery made this particular situation special. Even more striking, perhaps, was that the proofs submitted by the various groups represented three genuinely different mathematical approaches. The story of the Jones polynomial and its aftermath neatly illustrates the bridges that exist between science and mathematics and the unpredictable relationships between seemingly unrelated mathematical specialties.

Far from being the end of the excitement, the new invariants prompted new mathematical questions. Both the Polish mathematicians and Kenneth C. Millett and his colleagues went on to find several more independent polynomials that describe various aspects of knots. One of these is the so-called oriented polynomial, which includes information about how the knot sits in space. The oriented polynomial for a given knot diagram is the net result of assigning one value to a crossing if the strand passes over the intersecting strand and a different value when it crosses underneath, while going around the loop in some defined but arbitrary direction. The resulting expression, though powerful for distinguishing knots, is unfortunately considerably more complex than other known invariants. It gains its power at the expense of simplicity. For all but the simplest knots, the calculations are so involved that trying to solve them without a very powerful computer would be impractical.

Other polynomial invariants and methods for computing them have since emerged, but the Jones polynomial itself remains the key to the puzzle. It's more complex than its predecessor, the Alexander polynomial, but simpler than its successors. One reasonably straightforward procedure for computing the Jones polynomial is also easy to implement in a computer program. The difficulty is that the time needed to compute the invariant goes up exponentially with the number of crossings. This makes a knot with, say, 40 crossings almost impossible to check by computer. Whether a faster algorithm exists is an open question.

Furthermore, no one really understands what the new invariants mean geometrically. The Jones polynomial apparently encodes many kinds of existing data about knots, but in very strange ways. Many mathematicians are quite amazed that any one polynomial can detect so many different knot properties. Nevertheless, the new invariant has already turned out to be a useful tool. Using it, mathematicians were able to prove something that they had long suspected: a knot in which all the crossings alternate (like the weft in a woven fabric) is in its simplest possible form.

It seems clear that all these invariants are part of a still bigger picture that mathematicians barely glimpse. They know that none of the first 12,965 knots has a polynomial that equals 1: the polynomial of the unknot. But they also know that present theories can't distinguish certain classes of distinct knots. In attempting to unlock the secrets of the new polynomial invariants, mathematicians experiment by drawing lots of pictures and using hours and hours of computer time. They audition knots, looking for qualities that would focus attention on what the new invariants reveal and what they hide. To create elaborate structures to test, they carefully paste together relatively simple cases with special properties.

Experience, experiments, and intuition blend in the search for polynomial features that may reflect something observable in pictures of a knot. Like physicists who are trying to make sense of the particles and forces that make up the physical world, knot theorists are looking for something akin to a grand unified theory that would explain all invariants and all knots. They hope eventually to find a complete invariant that distinguishes any two knots.

Still missing, however, is the "bus-stop" invariant—a formula so simple that a mathematician waiting at a bus stop could pick any knot that springs to mind and quickly compute whether that particular tangled mess is really an unknot or a disguised version of some other familiar knot. The tougher problem of completely classifying knots using a simple set of easily computed invariants seems well beyond the reach of current methods. Meanwhile, encouraged by the unexpected links between seemingly unrelated mathematical pursuits, mathematicians continue to explore a variety of routes that may yet lead them to the supreme invariant.

Knot theory is also making a comeback in the sciences, not as Lord Kelvin's ethereal knotted vortices, but in the twisted, looping, stringlike structures of chemistry and molecular biology. Most famous of these structures is the long, skinny, twisted ladder of deoxyribonucleic acid (DNA) that embodies the genetic code governing life. A single DNA strand, if it had the width of a telephone cord, would be more than a mile long. Outside the cell, the strand looks like a tangled, disorderly heap of spaghetti. In the cell, the pair of helical DNA strands may be carefully folded or twisted into a springlike coil or joined end to end to form a loop. Two loops may intertwine as links in a chain. Strands may snake around into a knot. Photographs of protein-coated DNA molecules, taken through an electron microscope, clearly show the turns and crossings. One such configuration is shown in Figure 3.23.

Molecular biologists are starting to use knot theory to understand the different conformations that DNA can take on. Recent advances in knot theory are helping them see how DNA becomes knotted or linked during replication and recombination and how the enzymes that do the cutting and gluing must perform their functions. Biologists can track the sequence of steps in which one structure is gradually transformed into another during the basic, life-supporting processes that take place within cells.

Knot theory has allowed researchers at the University of Califor-

**F I G U R E   3.23**     A protein-coated loop of knotted DNA can be untwisted to reveal that the loop is a trefoil knot.

nia at Berkeley to make sense of the confusing array of differently configured DNA strands detected within a cell. It seemed unlikely that a single mechanism could account for the generation of the full variety of configurations observed. The researchers had to find out whether two strands of knotted DNA are really the same or different and how one knot is related to another—the same questions with which knot theorists have long wrestled. When the researchers arranged the molecules using a mathematical sequence of knots and links, they found a logical pathway. To test their idea, the researchers predicted that a specific six-crossing knot must be the next step, and when they went looking for it, they found the predicted strand (*see Figure 3.24*).

Chemists interested in synthesizing new compounds are also beginning to pay attention to topology and knot theory. In general, they try to create new, unusual molecules by changing the way atoms are connected. In this way, they hope to increase their understanding of how the chemical process of building compounds works. Geometry and especially topology suggest a range of targets for this effort. Inspired by the rigid figures of euclidean geometry, chemists in the last few years have managed to synthesize molecules such as tetrahedrane and dodecahedrane, in which carbon atoms are connected to form the vertices of a tetrahedron and dodecahedron.

Now that some chemists are beginning to think in terms of the infinitely flexible models of topology rather than the rigid ones of euclidean geometry, a new set of intriguing targets for synthesis has appeared, including knotted rings and large rings of atoms linked in a chain. Although several syntheses of molecular linked rings, termed *catenanes,* have been achieved in the last 25 years, no knotted ring has yet been prepared totally by chemical means. The difficulty lies in the unlikelihood of threading a chain of atoms through a loop to form a molecular half hitch. Chemical reactions joining the ends of long strings of atoms are now carried out routinely, but the probability of forming a knotted ring in this way is extremely small.

Hope for creating a tight molecular knot (unlike the looser, naturally generated knots in DNA) has received a boost from a novel

**FIGURE 3.24**    By appropriate cutting and splicing, an unknot can be turned into a six-crossing knot *(upper right)*. Such a knot was later found among strands of DNA *(bottom)*.

technique that relies on the properties of a Möbius strip. Cutting the strip in half lengthwise results not in two separate pieces but in a single strip with four half twists. Other cuts and conformations result in various twisted loops, knotted rings, and separate but linked rings. In 1981, David M. Walba of the University of Colorado managed to create a molecular Möbius strip with half a twist by joining the ends of a ladder-shaped, double-stranded strip of carbon and oxygen atoms. The technique of using molecular ladders may yet lead to the first chemical synthesis of a trefoil knot.

Walba's work, like that of molecular biologists working with the twists and turns of DNA, has itself suggested many new mathematical questions. In a crossover that enriches both pursuits, chemists and biologists are drawn into the fascinating mathematical world of knots and links as mathematicians tangle with molecules and chemistry.

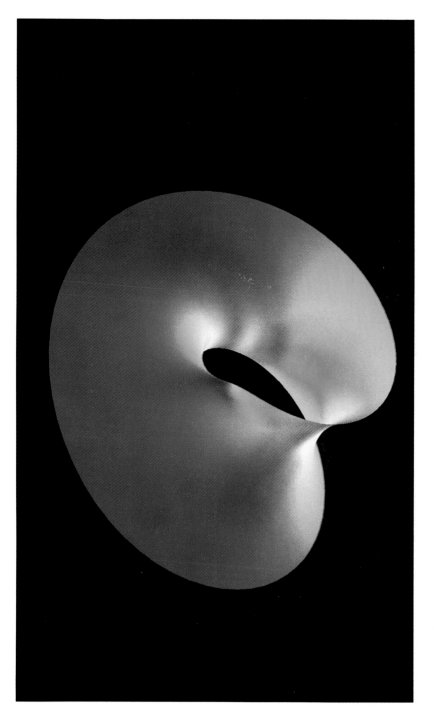

1   Minimal Möbius

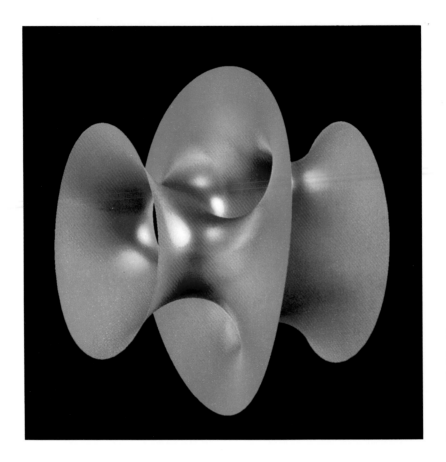

3    Peeling a Psychedelic Onion

4    Sliced Doughnuts

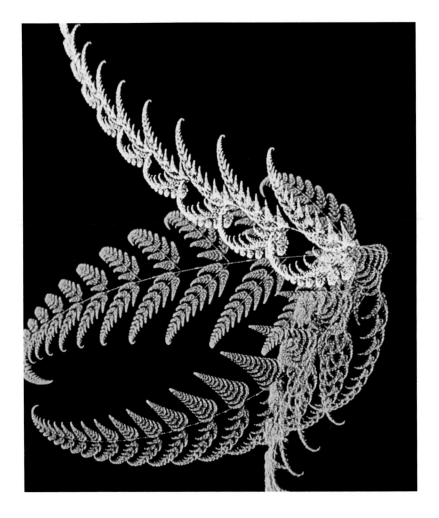

7  A Tree Grows in BASIC

8    Branching to Infinity

# 4

## SHADOWS
## FROM
## HIGHER
## DIMENSIONS

For many dabblers in geometry, the doorway to higher dimensions has been a slim volume called *Flatland*. The book's central figure and narrator, "A Square," takes his readers into a two-dimensional world where a race of rigid geometric forms live and love, work and play. Like shadows, the denizens of Flatland freely flit about on the surface of their world but lack the power to rise above or sink into it. All Flatland's creatures—straight lines, triangles, squares, pentagons, and higher figures—are trapped in their planar geometry *(Figure 4.1)*.

## FLATLAND AND BEYOND

On the surface, Edwin A. Abbott's narrative, written more than a century ago, appears simply to be a good story and a clever mathematics lesson. Although their real shapes are two-dimensional, Flatlanders appear to one another as straight lines. Residents of a three-dimensional world see the reason for a Flatlander's limited vista.

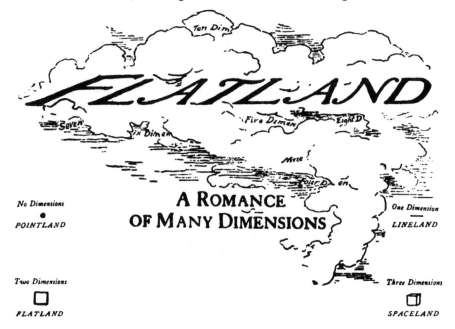

*"O day and night, but this is wondrous strange"*

No Dimensions
•
POINTLAND

A ROMANCE
OF MANY DIMENSIONS

One Dimension
—
LINELAND

Two Dimensions
FLATLAND

Three Dimensions
SPACELAND

**FIGURE 4.1** The title page from the first edition of *Flatland*, published in 1884, hints at the book's whimsical nature.

Viewed from above, a coin sitting on a table clearly looks circular. As the angle of view shifts closer to the plane of the table, the coin looks more like an oval. At Flatlander level, along the table's surface, the oval thins to nothing more than a straight line. Using such examples in *Flatland*, Abbott neatly introduces key ideas in projective geometry and illustrates other important mathematical concepts.

*Flatland* is also a pointed satire that reflects widely debated social issues of Victorian Britain. Abbott was a strong advocate of women's rights who couldn't resist taking a satirical swipe at his society's attitudes toward women. Flatland women are merely Straight Lines. Lower-class men are Isosceles Triangles; Squares make up the professional class; Nobles are polygons with six or more sides; and Priests, the highest-ranking members, are perfect Circles.

"[A] Woman is a needle; being, so to speak, all point, at least at the two extremities," says "A Square," the scholarly commentator. "Add to this the power of making herself invisible at will, and you will perceive that a Female, in Flatland, is a creature by no means to be trifled with." Nevertheless, Flatland women also are "devoid of brain-power, and have neither reflection, judgment nor forethought, and hardly any memory." In this planar world, men believe that educating women is wasted effort and that communication with women must be in a separate language that contains "irrational and emotional conceptions" not otherwise found in male vocabulary.

When he wrote *Flatland*, Abbott was headmaster at the City of London School, an institution that prepared middle-class boys for professional careers and for places at universities such as Cambridge. He produced dozens of books, including school textbooks, historical and biblical studies, theological novels, and a well-regarded Shakespearean grammar that strongly influenced the study of the Bard's plays. At first, *Flatland* seems out of place within this collection, but a closer look shows that it combines elements from Abbott's broad range of interests, from the nature of miracles to the reform of mathematics education. Abbott was part of a group of progressive educators who sought changes in the mathematics requirements for university entrance, which at that time included the memorization of lengthy proofs in euclidean geometry. Abbott's group considered such an exercise a waste of time and felt it narrowed the study of geometry unnecessarily.

Despite Abbott's views and the growth of popular interest in concepts of a fourth dimension and even higher dimensions, the conservative mathematics establishment in Britain prevailed. Establishment mathematicians refused to admit that higher-dimension geometries were even conceivable, maintaining that such concepts would call into question the very existence and permanence of mathematical truth, as represented by three-dimensional euclidean geometry. Abbott challenged the establishment and deliberately called Flatland's university "Wentbridge"—a sly dig at Cambridge.

*Flatland* also represents one of Abbott's attempts to reconcile scientific and religious ideas and to clarify the relationship between material proof and religious faith. In one episode, "A Square" receives a visit from a ghostly sphere, who tries to demonstrate to the bewildered Flatlander the existence of Spaceland and a higher dimension. The visiting sphere argues that he is a "Solid" made up of an infinite number of circles, varying in size from a point to a circle 13 inches across, stacked one on top of the other. In Flatland, only one of these circles is visible at any given time. Rising out of Flatland, the sphere looks like a circle that gets smaller and smaller until it finally dwindles to a point, then vanishes completely *(see Figure 4.2)*.

When that demonstration fails to persuade "A Square" that the sphere is truly three-dimensional, the visitor tries a more mathematical argument. A single point, being just a point, he insists, has only one terminal point. A moving point produces a straight line, which has two terminal points. A straight line moving at right angles to itself sweeps out a square with four terminal points. Those are all conceivable operations to a Flatlander. Inexorable mathematical logic forces the next step. If the numbers 1, 2, and 4 are in a geometric progression, then 8 follows. Lifting a square out of the plane of Flatland ought to produce something with eight terminal points. Spacelanders call it a cube. The argument opens a path to even higher dimensions.

Through mathematical analogy, Abbott sought to show that establishing scientific truth requires a leap of faith and that, conversely, miracles can be explained in terms that don't violate physical laws. Like early scientific theories, miracles could be merely shadows of phenomena beyond everyday experience or intrusions from higher dimensions.

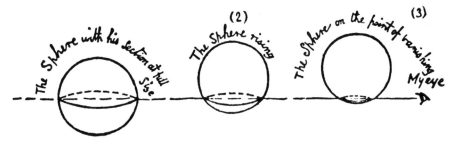

**FIGURE 4.2**    To a Flatlander, a Sphere passing through Flatland appears as a circle of changing diameter.

*Flatland* raises the fundamental question of how to deal with something transcendental, especially when recognizing that one will never be able to grasp its full nature and meaning. It's the kind of challenge that pure mathematicians face when they venture into higher dimensions.

How do mathematicians organize their insights? How do they see and understand multidimensional worlds? How do they communicate their insights? *Flatland* is a novel approach toward answering those questions.

## SHADOWS AND SLICES

Venturing into the fourth dimension can be as dramatic and rewarding as seeing the earth from space for the first time. The view from a higher dimension reveals features and patterns barely noticeable within the lower-dimensional world itself. A truly global picture emerges. Human beings, living in a three-dimensional world, easily recognize the triangles, squares, and other polygons of Flatland, whereas Flatlanders see only lines. In the same way, it takes a view from the fourth or a higher dimension to catch glimpses of the true nature of three-dimensional objects and to see their relationships to one another.

The mathematical journey into the fourth dimension starts with a point — a zero-dimensional object having no length, breadth, or height. A point stretches into a line, which in turn sweeps out a square, which then balloons into a cube. Once a line forms, at each succeeding stage, the figure expands in a new direction at right angles to those directions already defined. Each step takes the figure into a higher dimension. Adding a fourth direction at right angles to the other three sweeps a cube into the fourth dimension *(see Figure 4.3)*. The result is a *hypercube.*

What does a hypercube, or for that matter any other four-dimensional object, look like? One way to keep track of a multidimensional geometric figure is to use a coordinate system to locate the collection of points that make up the object. It takes a single number to specify the position of a point in one-dimensional space, just as a roadside signpost marks the distance from a certain town. Two numbers locate a spot in two dimensions. On the earth's two-dimensional, but curved, surface, longitude and latitude pin down any location. Similarly, $x$ and $y$ coordinates — a pair of numbers — can be used to specify each corner of a square drawn on a piece of paper. In three dimensions, everything is located by three numbers. Hence, a flight over the

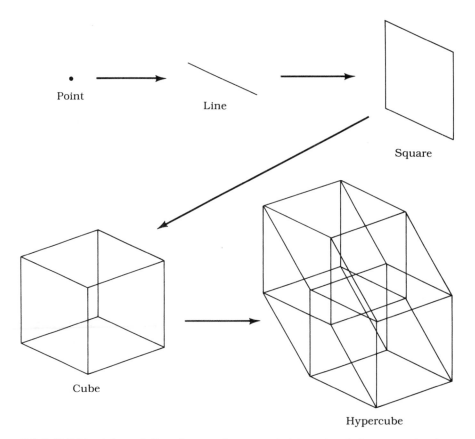

Point

Line

Square

Cube

Hypercube

**FIGURE 4.3** A line forms when a point is extended in a particular direction. Shifting the line at right angles to its length generates a square, whereas moving the square at right angles to its plane produces a cube. Moving the cube at right angles to all three previously defined directions creates a hypercube.

earth's surface requires an additional, third dimension—the altitude. Logically, four numbers or coordinates must then represent a point in four dimensions; five coordinates, a point in five-dimensional space; and so on.

However, most mathematicians need no pictures to step into higher-dimensional mathematical structures. They have long been able to deal abstractly with multidimensional problems by treating the concept of dimension more as a mathematical filing cabinet than as a direction in geometric space. Each filing folder in the cabinet represents a dimension in which all the information about one particular variable can be collected and stored. Problem solving and theorem proving are essentially manipulations, according to specific rules, of the numbers and symbols in these folders.

In fact, any set of four numbers, variables, or parameters can be considered either as a four-dimensional entity or as a string of numbers that can be filed or manipulated in certain ways. One of the most famous examples is embedded in Albert Einstein's theories of special and general relativity. In these theories, three-dimensional space and time together make up a four-dimensional continuum in which space and time are intimately interconnected.

Einstein's space-time is only one of many different types of four-dimensional spaces. A geologist may study past climates by correlating latitude and longitude with the amount of pollen found at various depths in bore holes drilled into the soil. The four variables—longitude, latitude, amount, and depth—add up to another kind of four-dimensional space.

Going beyond the fourth dimension is as easy as adding more variables. Some physicists have proposed modifications to Einstein's view of the universe. Their theories introduce seven or more dimensions in the form of miniature hyperbubbles hanging from each point of space-time. Applied mathematicians trying to solve the equations that would allow a business to manufacture and distribute its products most efficiently may deal with thousands of variables, from the cost of various raw materials to the distances between cities. The variables are nested in multidimensional spaces impossible to visualize.

A few mathematicians have over the years speculated about a fourth geometric, or spatial, dimension. In 1827, August F. Möbius noticed that the silhouette of a left hand can be turned into the silhouete of a right hand by passing the figure through the third dimension, one higher than the dimension of the figure itself. Anyone can duplicate his mental feat by using a cutout drawing of a hand. Picking up the cutout and flipping it upside down before returning it to a table transforms the figure into its mirror image. Perhaps, Möbius ruminated, a four-dimensional space would be needed to turn a three-dimensional, left-handed glove into a right-handed one.

Later in the nineteenth century, several scholars, including American mathematician W. L. Stringham, studied and published drawings of four-dimensional figures. Stringham's sketches in particular and the efforts of mathematician C. H. Hinton, who believed that people could be trained to see a hypercube, prompted a great deal of public interest. Even the prominent mathematician Henri Poincaré discussed the possibility of perceiving the fourth dimension.

Until the computer came along, attempts to picture otherworldly objects like the hypercube had limited success. Human beings had to cope with the same kind of problem that "A Square" faced when he encountered a sphere in Flatland. As "A Square" found, it's not easy to get a complete, meaningful view of a higher-dimensional object.

One possible solution is to look at the object's shadow; another is to picture slices through its body. Shadows and slices provide two distinct sets of views that reveal different kinds of information about a geometric object.

Consider the shadow cast by a three-dimensional chair. The chair is usually recognizable from its two-dimensional shadow alone. Although the shadow exhibits distorted lengths, skewed angles, and planes hidden by edges, it still preserves the connections and relationships between the chair's legs, back, and seat. In the same way, the purely mathematical operation of projection, akin to casting a shadow, preserves a figure's continuity, although the object's projection may be distorted.

Slices through an object provide a sequence of cross sections that preserve the object's angles and lengths, although the relationships among parts of the object may be ambiguous. For example, cutting along planes parallel to the floor gives a varying picture of a chair. Near the floor, a slice through the legs reveals four disconnected circles. Another slice may capture the square seat; yet another, the thin rectangle of the back. The mind's eye can reassemble these images into a picture of a chair.

A computer is able to manipulate and plot the four numbers or coordinates that locate each point in a four-dimensional object. But because it can display just two or three of the object's four dimensions, the viewer again sees only shadows or slices of the object. Nevertheless, a computer's speed and flexibility greatly extend the number and types of views of four-dimensional objects at which a mathematician can look.

The shadows cast by an ordinary, three-dimensional cube, transparent except at the edges, give some clues about what to expect from a hypercube. A cube's two-dimensional shadow can resemble a square within a square (see Figure 4.4). Rotating the cube produces a hypnotizing geometric minuet as the shadow's lines lengthen and contract and its squares diminish and expand. So, too, a rotating hypercube springs to life in a movie pieced together from a sequence of computer-generated images. At first, it may look like a square with 4 corners and 4 edges. That square turns out to be just one of 6 faces of a cube, which appear as parallelograms, sharing a total of 12 edges and 8 vertices. Further rotation shifts the figure to reveal a starburst of lines — the 8 ordinary cubic hyperfaces of a hypercube, which has 24 square faces, 32 edges, and 16 vertices (see Figure 4.5).

A similar grand tour of the hypercube seen in perspective is even

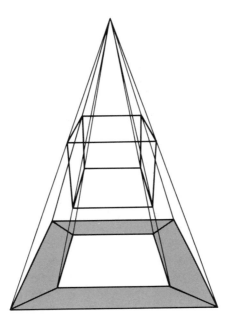

**FIGURE 4.4** A cube with opaque sides and a transparent top and bottom casts a shadow that looks like a square within a square.

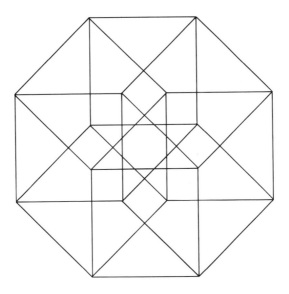

**FIGURE 4.5** This view of a hypercube shows the figure's 16 vertices and 32 edges.

more dramatic. If a hypercube is illuminated from a point lying outside normal, three-dimensional space, then its three-dimensional shadow looks like a small glass cube floating within a larger glass cube. As the hypercube rotates, the inner cube appears to shift, flatten, and turn inside out *(see Figure 4.6)*.

Slices through a cube hint at the hidden riches in store when slices of a hypercube are viewed. Cross sections of a cube may show up as triangles (when a corner is lopped off), as squares and assorted other quadrilaterals, and even as hexagons. In four dimensions, slicing with a three-dimensional knife produces even stranger shapes: distorted cubes, various prisms, and complex regular and semiregular polyhedra, often in unexpected arrangements.

Shadows and slices of higher-dimensional objects provide a useful way for mathematicians to visualize these strange forms. Just as experienced musicians can look at a musical score and imagine the sounds, mathematicians accustomed to working with austere equations and stingy notation can read their ordered symbols and painstakingly construct a mental image of what they have composed. But the mental image isn't always enough. Vivid pictures, like brilliant musical performances—in which the sonorous harmonies and subtle patterns written on paper spring to life—bring new richness and beauty to mathematics, which mathematicians and nonmathematicians alike can appreciate.

One mathematician who has recently spent a great deal of time in the shadowy world of computer-generated images of four- and higher-dimensional objects is Thomas Banchoff of Brown University. With computer graphics, he and other research mathematicians can turn the study of geometric phenomena into an observational science. To Banchoff, grand tours of visual images suggest relationships and conjectures that don't arise easily just from equations or lists of data points. In his words: "You can see things before you go ahead and prove them."

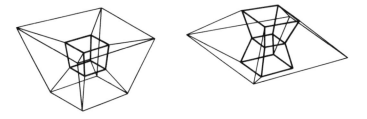

**F I G U R E   4.6**    Two views of a hypercube rotating in space demonstrate how different the hypercube's three-dimensional shadow may look from different angles.

Banchoff and his colleagues have created several movies that vividly show off the startling properties of hypercubes, hyperspheres, and other higher-dimensional forms, and allow mathematicians to explore the curious properties of these figures. Their insights are useful in mathematical fields such as differential geometry (minimal surfaces in higher dimensions and curved spaces), topology, and data analysis.

The various ways of plotting and manipulating four-dimensional data with computer graphics are handy tools for many kinds of data analysis. Often, more than three different types of measurements are necessary to characterize a physical situation. For example, U. S. Navy personnel are interested in uncovering the relationship between the temperature, salinity, biomass, and oceanic current strength at particular points in the ocean so that they can track sound waves traveling through water. Computer-generated pictures allow them to search for patterns that help them understand how these various physical factors are related.

A computer-generated sequence of pictures can also hold a surprising amount of information. Each tiny dot — called a picture element, or pixel — on a video monitor has a specific location that can be specified by two coordinates representing two variables. That pixel may also have a color, which could represent a third variable. Animation adds a fourth dimension, usually time. In this way, an astrophysicist, say, can view slices of a complex event such as the collision of two neutron stars, keeping track of the event in a four-dimensional space of position (two coordinates), density, and time.

Playing with color and relying on various visual illusions to create the impression of height and depth add to the number of dimensions or variables that can be represented in a computer-generated image. That's useful in an age when researchers often have to cope with many variables and billions of pieces of data that otherwise would fill reams of computer paper.

## A GLOBAL VIEW

From the graceful, flowing forms of minimal surfaces to the starburst of edges in a hypercube, many striking computer-generated mathematical images are coming out of exploratory forays into the realm of geometric shapes in three- and four-dimensional space (three-space and four-space, for short). A particularly rich source of images is the bizarre world of three-manifolds in four-space.

When topologists started to explore the world of geometric forms in higher dimensions, they found that neither the intuition nor the vocabulary of ordinary geometry was sufficient to describe and classify the new forms they were discovering. They introduced *manifold* as a general term for describing a certain common type of higher-dimensional, geometric object.

Manifolds are surfaces and shapes, sometimes very complex, that appear to be euclidean when a small region is examined. On a large scale, however, these forms fail to follow the rules for a euclidean geometry. Roughly speaking, the earth's surface is an example of a manifold. To a gardener marking off a square plot in a field, the earth's curvature is barely noticeable, if at all. The gardener doesn't have to worry about opposite borders that aren't quite parallel or corners that aren't quite right angles.

A circle, although it curves through two dimensions, is an example of a one-dimensional manifold, or one-manifold. A close-up view reveals that any small segment of the circle is practically indistinguishable from a straight line. Similarly, a sphere's two-dimensional surface, even though it curves through three dimensions, is an example of a two-manifold. Seen locally, the surface appears to be flat.

Picturing a three-manifold is somewhat more difficult. Standing in one spot and looking around within a three-manifold wouldn't tell an observer much about the manifold's shape. It would look like ordinary three-dimensional space. Indeed, just as it's hard to imagine the two-dimensional surface of a sphere without putting it into three-space, a three-manifold is most easily viewed from four-space.

The theory of manifolds arose in the nineteenth century out of a need to understand solutions to sets of equations expressing relationships between two or more variables. For instance, all the possible solutions to a two-variable equation can be plotted as a set of points in the plane. Each point represents a pair of values that make the equation true. Those points typically fall on a curve. Similarly, the set of solutions to an equation that has three variables can be plotted as a two-dimensional surface in three-dimensional space. The points could, for example, fall on the surface of a sphere or some other two-manifold. For an equation with more than three variables, the set of solutions may look like a multidimensional manifold embedded in a space one step higher in dimension than the manifold itself.

It takes considerable mental agility to picture even a simple three-manifold, such as a three-dimensional sphere. That contorted object can be thought of as the result of gluing together the skins of two balls so that one sphere is turned inside out over the other. Astonishingly, the mathematics of three-manifolds turns out to be exceedingly intricate and more difficult than the mathematics of manifolds in any other dimension.

The place to start on a trip into this strange world is back in everyday three-space — the three-dimensional, euclidean space familiar to all of us. The two-sphere is a convenient starting point. Assigned a radius of 1 unit, it has a remarkably straightforward algebraic formula: $x_1{}^2 + x_2{}^2 + x_3{}^2 = 1$, where the coordinates $x_1$, $x_2$, and $x_3$ pinpoint the location of each point on the sphere with respect to three axes at right angles to one another.

To be seen, the surface of a sphere ought to be drawn or represented in three-dimensional space, just as a globe represents the earth's features. But it's also possible to represent the surface of a sphere in two dimensions: on a flat surface or on the screen of a video monitor. Mapmakers do this all the time, having developed a variety of ways to map continents and oceans onto pages in an atlas.

The standard stereographic projection is one means of achieving the transformation from a globe's surface onto a flat map. Every point on the sphere ends up at a particular spot on the planar map. If the north pole happens to be at the point (0, 0, 1), that is, one unit up the $x_3$ axis, then the projection maps every point on the sphere onto a plane, defined by the $x_1$ and $x_2$ axes. That plane slices through the globe along its equator. All the sphere's points, except the north pole, end up somewhere in the plane (*see Figure 4.7*).

A simple mathematical formula gives every starting point on the sphere its particular destination on the plane. A point on the globe's equator stays on the equator, now seen as a circle on the planar map.

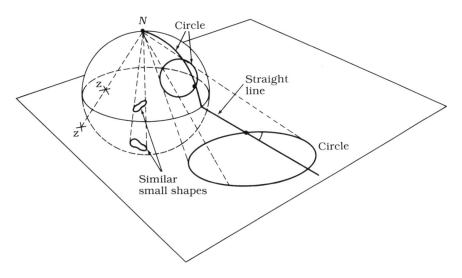

**FIGURE 4.7** The stereographic projection of a sphere onto a plane preserves the shapes of some geometric forms and alters others.

The point $(\frac{3}{5}, 0, \frac{4}{5})$ in the northern hemisphere is sent to the spot $(3, 0)$. In fact, every point in that hemisphere ends up somewhere on the plane outside of the circle marked by the equator. A point, such as $(\frac{3}{5}, 0, -\frac{4}{5})$, in the southern hemisphere goes to a planar point $(\frac{1}{3}, 0)$, that falls inside the equatorial circle. In general, each point $(x_1, x_2, x_3)$ on the sphere arrives at $\left(\dfrac{x_1}{1-x_3}, \dfrac{x_2}{1-x_3}\right)$ in the equatorial plane. The south pole sits at the center $(0, 0)$, while the north pole is banished to infinity.

Although the resulting map is a decidedly distorted view of the globe, with every southern point crowded inside a circle and all the northern points spread out over the rest of the plane, no point on the sphere has been lost and no points overlap. Structures and patterns visible on a globe appear in warped but recognizable forms on the resulting map. By carefully studying what happens to the shape of particular global structures when they are transferred to a flat surface, mathematicians can learn to deduce or construct global properties merely by examining features on a flat map. For four-dimensional objects, which can't be seen directly, mathematicians have to rely entirely on three-dimensional "maps" to get a sense of their four-dimensional counterparts.

One way to see what's happening under a stereographic projection is to imagine a transparent globe on which circles of latitude have been drawn. If a light is fixed at the usual position of the north pole and the globe's south pole rests on a flat surface, then the shadow cast by the circles of latitude is a set of concentric circles. If the globe is tilted while the light stays fixed in position, the images of the circles shift and become distorted. Any circles that momentarily happen to pass through the light's position would cast a fleeting straight-line shadow.

Thus, any circle on the two-sphere reappears as a circle in the plane. The only exceptions are circles that happen to pass through the north pole. They project to a straight line. Thus, a set of stripes, all of equal width and parallel to circles of latitude, would project into a bull's-eye of concentric rings on the plane. Each ring gets wider as its distance from the center of the plane, or south pole, increases *(see Figure 4.8)*.

The procedure used to create planar images of the two-sphere can be extended to generate images of a hypersphere, or three-sphere, as it would appear in three-dimensional space. The three-sphere in four-dimensional euclidean space is defined by the equation $x_1^2 + x_2^2 + x_3^2 + x_4^2 = 1$. Compared with the two-sphere, all that's new is the addition of the coordinate $x_4$.

What does the three-sphere look like? The experience of "A

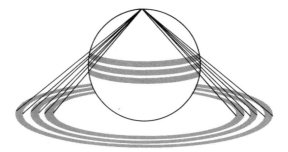

**FIGURE 4.8** Projections of bands encircling a sphere onto a plane.

Square" in Flatland suggests one way to visualize the object. When Flatlanders view a sphere descending from above into their planar world, they see only the part of the sphere that intersects their plane. Their first view is of a point, followed by circles of increasing diameter up to the largest circle, when the sphere's equator passes through Flatland. Then diminishing circles appear until they dwindle to a point and vanish. By analogy, a three-dimensional human visited by a hypersphere would first see a single spot, a tiny sphere, growing steadily through a sequence of ever larger spheres. Then the spheres would begin to shrink, at last disappearing altogether.

Projections, such as the four-space equivalent of the stereographic projection, provide some mathematically more useful glimpses of a hypersphere. One example of such a projection is the Hopf map, named for German mathematician Heinz Hopf. This map takes points on the three-sphere and systematically finds places for them on the two-sphere. Mathematically, each point $(X_1, X_2, X_3, X_4)$ on the three-sphere becomes the point $(x_1, x_2, x_3)$ on the two-sphere, where

$$x_1 = 2 (X_1X_2 + X_3X_4),$$
$$x_2 = 2(X_1X_4 - X_2X_3),$$
$$x_3 = (X_1{}^2 + X_3{}^2) - (X_2{}^2 + X_4{}^2)$$

Under the Hopf map, every point on the two-sphere represents a circle (called a Hopf circle) on the three-sphere. Reversing the projection by going from the two-sphere to the three-sphere, that is, finding the map's inverse, reveals that a circle of latitude on the two-sphere represents a doughnut-shaped surface, or torus, in the three-sphere. This is enough to build up a picture of the hypersphere. Just as two points (representing the north and south poles) and a series of adjacent bands on a planar map can be used to build up a complete picture of a two-sphere, two linked circles and a family of tori make up a full three-sphere. A sliced three-sphere can be imagined as a sequence of doughnuts within doughnuts, with each successive doughnut swelling in size as its distance from the center increases *(see Figure 4.9)*.

By manipulating the computer-generated image of a three-sphere, mathematicians can turn this abstract mathematical object into the star of an animated film and use it to explore properties of three-manifolds. By tinkering with the equations that define these objects, they can highlight particular surface features. They can slice the objects open to get a better view. They can watch the objects move, linking strings of images to reveal coherent patterns, such as the smoothly twisted tori that swell up and sweep past one another when a hypersphere rotates *(see Color Plate 4)*.

Pictures alone, however, don't tell the whole story. The Hopf map itself was the first example of a special type of projection of points from one sphere to another sphere of lower dimension. Its existence inspired further mathematical research in various geometric topics, including the sorting out of several types of three-manifolds in four-space.

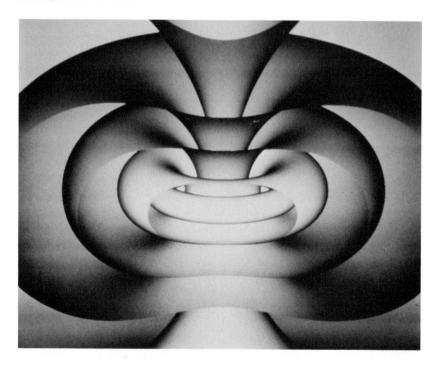

**F I G U R E   4.9**    A view of a hypersphere with its image sliced open to expose its inner structure.

Three-manifolds, such as the hypersphere, show up in applied mathematics and physics. Different sets of mathematical equations, which may be useful for describing physical processes as diverse as fluid flow and crystal growth, generate different curves and shapes in space. The changing behavior of complex systems over time turns out to correspond neatly to motion on the curving shapes of certain surfaces. In many instances, understanding the solutions of particular equations can be turned into a question of geometry. Although some equation solutions show up as collections of points on a sphere or a torus, others may reside on a more exotic object, such as a one-sided, one-edged Möbius strip or a three-dimensional tube called a Klein bottle, whose inside surface loops back on itself to merge wih the outside.

The hypersphere, for one, makes an appearance in the analysis of motion in phase space, in which each dimension represents one of the variables in the equation used to model a particular system. For a mechanical system, the variables may be just the positions and velocities of each particle in the system. Beginning at a point representing the initial values of all the variables, the equation generates a trajectory that winds through phase space. The location of a point on the trajectory at any time contains all the information needed to describe the state of the system — that is, the values of all the variables — at that particular time. Although the motion itself isn't directly visible, such phase-space portraits provide a useful collection of information about the motion.

For example, in the motion of a pair of oscillators, such as two windshield wipers, it takes one variable, the angle, to define each wiper's position and another variable to define each wiper's velocity. Together, the positions and velocities of both wipers trace out a path on the surface of a hypersphere in four-dimensional space. A similar analysis applies to a pendulum bob fixed to a stiff rod and mounted so that the bob is physically free to trace out a path on the surface of a sphere. It takes two coordinates to define the bob's position and two to specify how quickly it is moving. Tracking the pendulum's path through phase space — that is, plotting its position and velocity at different times — gives a graphical history of the pendulum's behavior. Four-space, however, has so much room that this curving, four-dimensional path could wander almost anywhere. Luckily, physical constraints — the laws of conservation of energy and momentum — keep the system within strictly defined bounds. The four-dimensional phase-space curves are confined to a three-manifold in four-space.

In normal situations, these curves, as seen in four-dimensional phase space, fall on the surface of a doughnut, or torus. Each torus represents a certain combination of energy and momentum, equivalent to the initial push given to a pendulum to get it started. In a hypersphere picture, the Hopf tori represent the constant "energy-momentum" surfaces, and the Hopf circles on them are curves representing the changing positions and velocities of each pendulum *(see Figure 4.10)*. These curves are essentially solutions to differential equations, called Hamilton's equations, which define how the motion changes with time. We take a closer look at the behavior of differential equations in Chapter 6.

Computer graphics provides a way for researchers to study what the solution curves look like for various physical systems and sets of initial conditions. These pictures greatly enrich the study of dynamical systems — the way things change in time or space.

## THE CLASSIFICATION GAME

Like botanists, mathematicians suffer from the urge to classify. Unlike botanists, who can never be sure that every plant species on earth has been collected and categorized, mathematicians can sometimes prove that a classification scheme for a given set of mathematical objects is complete. They can formulate an ironclad guarantee

**FIGURE 4.10**    Two linked toroidal energy-momentum surfaces for a pair of linear harmonic oscillators.

essentially stating that no one will ever discover an object that belongs to the set yet doesn't fit one of the mandated categories. What keeps the game interesting, however, is that such proofs are often elusive. To mathematicians, success or failure in these classification efforts serves as a test of the power of their mathematical methods.

One of the greatest classification quests in mathematics, which still isn't over, is the search for a complete classification of geometric shapes. The task was relatively easy centuries ago, in the days when the family of known geometric forms encompassed familiar, well-defined, rigid figures: lines and circles, triangles and cubes, assorted polygons and polyhedra. Then topology reared its rubbery head, and flexibility became the rule. On top of that, the spread of geometry into higher dimensions, from curves and surfaces into manifolds, left mathematicians with an unruly zoo of geometric forms. Attempts to classify these myriad shapes tended to lag behind discoveries of new ones. It was as if a botanist, not yet finished with the earth, were faced with a rapid succession of new worlds of exotic species, all defying easy classification.

One of the major triumphs of nineteenth-century mathematics, and of topology in particular, was the complete classification of two-dimensional surfaces (two-manifolds). As outlined in the previous chapter, all surfaces, if they are treated like plasticene sheets, can be bent, stretched, curved, or deformed—so long as they are not torn or glued—until they match a set of basic forms. Every conceivable surface, from the skin of a doughnut to the convoluted exterior of a piece of coral, is equivalent to a sphere with a certain number of attached handles and a certain number of "ends," or punctures, that penetrate its skin. Because the number of handles (or genus) of a surface never changes no matter how the surface is deformed, this quantity stands as a way to classify surfaces. Hence, the surface of a coffee cup and the skin of a doughnut, having the same genus, fit in the same category: a one-handled, or one-holed, sphere.

Three-manifolds aren't quite so simple to classify. They're more difficult to visualize, and four-space allows plenty of room for their strange, complex forms. A human being in three-space would have the same trouble that a nearsighted ant crawling on the surface of a rather large doughnut would have in telling whether the doughnut has a hole. The ant can't stand outside the surface and see the hole, and it can't step off the surface to fall into it. The hole can be seen clearly only from a higher dimension. Similarly, a human explorer wandering around on a three-manifold wouldn't see much out of the ordinary. It needs a vantage point in four-space to detect holes and other features of a three-manifold.

However, we can try some indirect approaches to the hole-detection problem. An inquisitive ant, wondering whether its surface is a sphere, torus, or some other exotic form, could detect a hole by taking circular tours around the surface. The ant carefully notes the properties of these closed loops as they shrink in size. Sometimes, the ant's path will form a closed loop that happens to thread its way through a hole. In that case, the ant would find that its circular path can't be shrunk beyond a certain point; it gets snagged on the hole. Loops that don't thread a hole, however, can be shrunk. If an ant finds that all possible closed loops on the surface can be shrunk, then it knows it's on a sphere.

The ant's experiment can be turned into a useful mathematical procedure for detecting holes. This procedure forms the basis for an algebraic method, developed by Henri Poincaré at the turn of the century, to detect and study the pattern of holes in a topological surface. His strategy, like the ant's, depends on the shrinking properties of loops on a given surface.

All closed paths on a sphere can be continuously shrunk to a single point. That sphere, then, is said to be simply connected. In contrast, there are fundamentally different types of paths on a torus, only some of which can be shrunk to a point *(see Figure 4.11)*. Those winding through the center hole differ from those winding around the center hole. And both are different from the cyclic paths that loop through the center as they go around the torus. Every surface other than a sphere contains some unshrinkable loops.

Poincaré expressed his scheme in terms of an algebraic structure called a group. In this context, a group is a kind of exclusive club whose members obey a set of strict mathematical rules. It's also a way

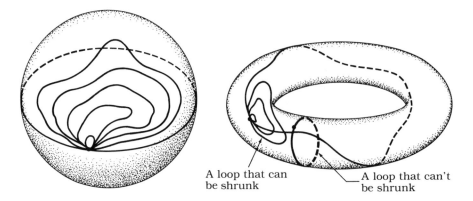

A loop that can be shrunk

A loop that can't be shrunk

**FIGURE  4.11**    Loops  on  a  sphere  and  a  torus:  some  loops  can  be shrunk to a point; others can't.

of expressing compactly certain characteristics of mathematical objects; in this case, the shrinking properties of loops on surfaces. Poincaré's fundamental groups specify the way in which closed loops on a manifold may be deformed into one another.

For a two-sphere, the fundamental group consists of a single member because any loop lying on a sphere may be tightened like a noose until it is shrunk to a point. For this and other reasons, topologists consider the two-sphere to be the simplest of all surfaces — the one with the fewest topological complications.

A torus requires two winding numbers, which represent the number of times a curve winds around the body of the torus (like the stripes on a candy cane) and the number of times a curve winds around the hole in the middle. Hence, the fundamental group for a torus consists of a pair of generators for representing the two possible winding numbers.

This means that the two-sphere and the torus are associated with different groups, reflecting the fact that the two surfaces are topologically distinct. One can't be deformed into the other. In this way, all possible two-manifolds can be labeled, and every surface fits into a particular category given by its fundamental group. That labeling matches the classification scheme based on the number of handles, or holes, a surface has. For a complicated shape, calculating the Poincaré group provides a handy way of classifying the object.

One consequence of this scheme is that for two-manifolds, any surface that happens to have a one-member, or trivial (as mathematicians like to call it), fundamental group must be topologically equivalent to a two-sphere. A three-dimensional sphere also turns out to have a trivial fundamental group. Considering the family resemblance between two-spheres and three-spheres, Poincaré conjectured that three-dimensional spheres behave just like two-dimensional ones. Any three-manifold that is associated with a trivial fundamental group should be deformable into a three-sphere. That is, any three-manifold that behaves like a three-sphere is a three-sphere. There are no fake spheres.

The Poincaré conjecture, still unproved despite numerous heroic efforts over the last few decades, ranks as one of the most challenging and baffling unsolved problems in mathematics. Whole fields of mathematics have been developed in the course of testing, on a case-by-case basis, not only the conjecture itself but also the extension of the conjecture to manifolds in higher dimensions.

The central problem in proving the Poincaré conjecture is that, unlike the two-manifold case, no one knows a classification scheme for three-manifolds. There is no list of all possible surfaces that can be checked one by one to make sure that all nonspheres contain unshrinkable loops. Even worse, attempts to classify three-manifolds seem to depend on knowing that the Poincaré conjecture is true. The only way out is to develop totally new methods that bypass the unsolved classification problem.

Ironically, the first major advances toward proving the Poincaré conjecture occurred not for three-manifolds but for higher-dimensional manifolds. One important tool for the study of higher-dimensional manifolds, developed in the early 1960s, was the technique of *surgery*: systematically studying the effect of removing certain pieces of a manifold, then gluing them back with a specific twist *(see Figure 4.12)*. In 1961, Stephen Smale developed and used just such a surgical method to break manifolds into digestible, bite-size pieces that could be easily moved around. His work led to a far-reaching theorem, one of whose consequences was the truth of the Poincaré conjecture for five- and higher-dimensional manifolds.

These higher-dimensional manifolds turn out to be easier to classify than three- or four-manifolds, but even spheres, the lowliest of

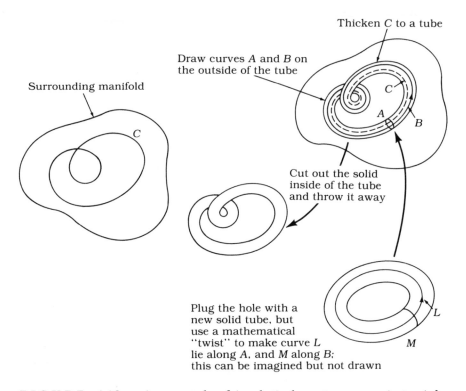

Thicken $C$ to a tube

Draw curves $A$ and $B$ on the outside of the tube

Surrounding manifold

$C$

$C$

$A$

$B$

Cut out the solid inside of the tube and throw it away

Plug the hole with a new solid tube, but use a mathematical "twist" to make curve $L$ lie along $A$, and $M$ along $B$: this can be imagined but not drawn

$L$

$M$

**FIGURE 4.12**    An example of topological surgery: removing a tube and then gluing a new one back in with a "twist."

higher-dimensional manifolds, sometimes display unexpectedly bizarre properties. For example, John Milnor in 1959, doing a kind of multidimensional plumbing—cutting holes here and there, connecting them by various tubes, then deforming the result into a new kind of sphere—stunned the mathematical world by showing that the seven-dimensional sphere can be made into a smooth, or "differentiable," manifold in 28 different ways. Milnor's result means that Poincaré's scheme of representing the solutions to differential equations on manifolds would, in dimension seven, encounter 28 different versions of the sphere.

After a spurt in the 1960s, efforts aimed at proving Poincaré's conjecture stalled. Two cases remained: three-manifolds and four-manifolds. That left mathematicians with the feeling that something unusual happens in spaces of dimensions four and five. In both, the number of dimensions is high enough to allow complicated behavior, but too small to leave sufficient room for mathematical maneuvering to eliminate those complications.

Four-manifolds finally fell into place in 1982. Using a system of *flexible handles*, Michael Freedman of the University of California at San Diego blended loops, surgery, and boundaries to solve several major problems in four-dimensional topology and to give a complete classification of all simply connected, or spherelike, four-dimensional manifolds. His basic idea was to start with a four-dimensional disk and build up a given four-manifold by adding handles one by one. One consequence was a proof of the Poincaré conjecture for four-manifolds. In addition, Freedman's extraordinary work unearthed several previously unknown examples of four-manifolds and many unsuspected connections between known manifolds.

In proving the conjecture, Freedman also took a big step toward classifying all four-manifolds. These manifolds fall into two types. *Differentiable manifolds* change smoothly, one feature blending into the next. *Topological manifolds*, a broader category that encompasses differentiable manifolds, also include surfaces that sometimes resemble earthquake-torn globes, having tortuous, jagged landscapes that change abruptly from place to place. Freedman proved that although every differentiable manifold is also a topological manifold, not all topological manifolds are differentiable. This phenomenon first occurs in four dimensions.

Furthermore, Freedman constructed a particular topological four-dimensional manifold called E8, which has eight two-dimensional holes and is so convoluted that it can't be smoothed. Freedman's proof that such a manifold exists led to the astonishing discovery, by Oxford mathematician Simon Donaldson, of a new four-dimensional space. Donaldson's result sharpened the view of many mathematicians that four-dimensional geometry is truly more interesting and complicated than initially they had any right to think.

What makes Donaldson's surprising discovery particularly intriguing is that it concerns four-dimensional space—a space that is clearly of physical importance and a space that is not particularly abstract. Physicists now suddenly face the mind-boggling prospect of having more than one mathematical way to construct what they conceive of as ordinary four-dimensional space-time. The geometric rules governing the two four-spaces are the same, but their structures are different.

Perhaps this new four-dimensional structure has no physical meaning, or perhaps it means that the universe we inhabit has a more peculiar structure than anyone ever imagined. In fact, the universe could be as convoluted as a tangled loop of string. Or there may be space-time universes, organized in different ways but coexisting with ours, with which we can't communicate in any way. Donaldson's startling mathematical result has provided physicists with much food for thought.

Freedman's work left just the original Poincaré conjecture unsolved—the case for three-minifolds. Mathematical surgery suggests why the three-manifold case is so complex. Surgery makes it possible to construct a three-manifold from any tangled loop of string, no matter how knotted or convoluted the tangle. As in knot theory, that's like confronting two separately snarled masses of fishing line and trying to determine whether or not the two piles of line are tangled in the same way. Unless it's possible to classify such tangles of line in a systematic way, there seems to be no hope that three-manifolds can be analyzed either. Solving one problem would solve the other.

Yet even the three-manifold case may be within reach. In 1982, Princeton University's William P. Thurston set forth a program that could lead to the classification of three-dimensional manifolds and a resolution of the original Poincaré conjecture. Thurston's approach uses surgery to reduce complex manifolds to simple cases in much the same way that factoring reduces large composite numbers to products of primes.

Thurston found a geometric pattern that may turn out to encompass all possible three-manifolds. The pattern involves a surprising connection between three-manifolds and the curved spaces of noneuclidean geometry. A century ago, topologists believed that only two-manifolds can be described as being flat (having zero curvature), spherical (positively curved), or saddle-shaped (hyperbolic or negatively curved). Three-manifolds look too complicated to fit into such simple categories. Unlike a two-manifold, which can be specified and listed according to its genus, every three-manifold, like a tangled loop of string, seems to have its own distinctive properties and to resist fitting into any larger pattern.

The example of two-manifolds suggests possible ways to tame the unruly three-manifolds. Two-manifolds can be represented as many-sided, or polygonal, figures whose edges are glued together in a particular way. For instance, a Flatlander moving on a square may find that when he moves off the top edge of the square, he reappears at the bottom. When the creature moves off the right edge, it reappears at the left. Many video games operate on the same principle to keep a figure on the screen. Such movements can occur if, in some sense, the top of the square is glued to the bottom, and the right edge to the left. What does the resulting surface look like? The first gluing creates a cylinder; the second curves the cylinder into a doughnut. The doughnut and the square topologically count as the same abstract manifold: a two-torus *(see Figure 4.13)*.

This gluing trick turns out to be useful in bringing geometry back into the study of three-manifolds, making many manifolds more easy to visualize. A three-manifold, for example, can be generated from a rectangular block, such as the space inside a room. It's a matter of abstractly gluing the front wall to the back wall, the left wall to the right wall, and the floor to the ceiling. If we actually made these gluings, the result would be a room that has to bend around and join itself in the fourth dimension. But all that is needed for the description of the manifold is given by the procedure for abstract gluing. A human being living in such a house would probably find its geometry quite disconcerting—walking through one wall and ending up on the opposite side of the room, going through the ceiling to come up through the floor, disappearing at one wall and reappearing at another. Because such motion is strikingly similar to a Flatlander's motion on a two-torus, this particular three-manifold is called a three-torus. Abstract gluing shows the close tie between curved three-manifolds and polyhedra such as the rectangular block.

However, at least two major differences hold between the geometry of two-manifolds and the geometry of three-manifolds. First, whereas two-manifolds can have three types of local geometry (flat, spherical, and hyperbolic), three-manifolds can be given five additional types of local geometric structure. The additional geometries are conceivable because in three or more dimensions a curvature is defined for each two-dimensional slice that passes through a point. Second, it is possible to combine three-manifolds in such a way as to yield new three-manifolds that cannot be given a locally homogeneous geometry. Fortunately, topologists know how to split a three-manifold into primitive pieces by purely topological methods.

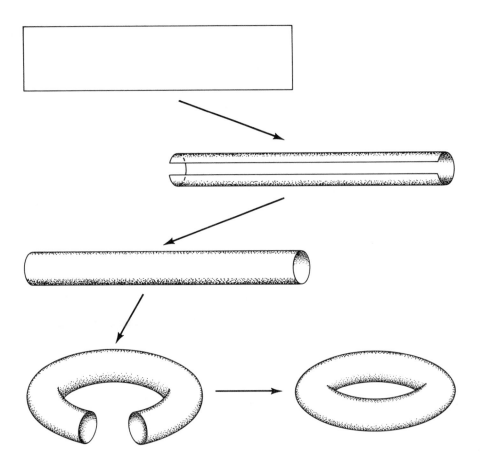

**FIGURE 4.13**    A torus is topologically equivalent to a rectangle: gluing the opposite edges of a rectangle turns it into a cylinder; joining the free pair of edges turns the cylinder into a torus.

In 1980, Thurston proved that most three-dimensional shapes fit into the hyperbolic category and that many exceptions fit into one of seven other, rarer classes of geometric forms. He conjectured that by looking at small pieces or neighborhoods within a manifold, topologists can put each piece of the manifold into one of eight categories, based on the geometric shape of the neighborhood. It's like having eight basic sets of clothing that fit anybody in the world. In a sense, mathematicians impose a geometric structure on a manifold by cutting it into a polyhedron that can then be abstractly glued back together in one of the eight basic geometric types. The trick is to prove that the eight geometric categories suit all conceivable three-manifolds.

Thurston's conjecture has been proved for wide classes of three-manifolds, and it has been tested by experiment on many other examples, either by hand or with the aid of a computer. So far, Thurston's scheme has never failed, and many mathematicians think it's unlikely a counterexample will be found. But no one knows yet for sure.

Thurston's insight has also brought some rigidity back into topology. Normally, topologists study properties that don't change when a shape is deformed. Thurston showed that sometimes it helps to impose a rigid geometric form to pin down topological structures. It makes classification, at least for three-manifolds, easier.

The lack of a proof for the Poincaré conjecture hasn't stopped mathematicians, however. They work around the conjecture by scrupulously avoiding hypotheses that depend on the truth of the Poincaré conjecture. Such caution is typical of mathematicians. Although most believe that the conjecture is true, a definitive proof is necessary before it can become part of the structure of mathematics. Moreover, as a result of this caution, when a proof finally comes, it won't directly affect much of the work topologists have already done. But the proof itself will likely involve new mathematical methods that turn out to be useful elsewhere.

## ———— LINEAR GYMNASTICS ————

The Too-Good-To-Be-True Bakery is famous for its best-selling bran muffins and carrot cakes. The shop sells a dozen muffins at a profit of $2.50 and a carrot cake at a profit of $1.50. With high demand and an unlimited supply of ingredients, the baker's choice is clear. Baking more muffins means earning a greater profit. One day, the baker notices that she has only a limited amount of flour on hand. Eggs are scarce. Her supply of honey is about to dry up. Because the recipes for bran muffins and carrot cake call for different proportions of these ingredients, the path to maximum profit is no longer quite so clear-cut. A dozen bran muffins require more flour than a carrot cake. Baking only muffins would use up all the flour and may let too much of the other ingredients go to waste. Perhaps baking a combination of both muffins and cakes would use up more of the ingredients and net a higher return.

The baker's plight is a simple, hypothetical example of a type of mathematical problem that involves finding the best solution among many possible answers. The mathematical techniques invented to solve such problems allow managers and planners to establish optimal strategies for allocating time, money, or supplies to keep costs down, profits up, or projects on schedule. To use these techniques, planners must be able to express the main features of their real-world problems in mathematical terms, without overlooking essential details and without excessively distorting reality.

Linear programming (where "programming" means planning) was invented to make such choices easier. "Linear" refers to the type of equations used to express the constraints bracketing a given situation. In making muffins, for example, the amount of each ingredient needed is assumed to be proportional to the number of muffins produced. Doubling the number of muffins doubles the amount of flour consumed; it also doubles the profit from the sale of muffins.

In the bakery problem, suppose that it takes 2 cups of flour to make a dozen muffins and only 1 cup to make a carrot cake. The total amount of flour available, 50 cups, limits the number of muffins and cakes that can be produced. That constraint can be expressed as an equation in terms of two variables. If $B$ represents the dozens of bran muffins to be produced and $C$ the number of carrot cakes, the constraint equation is simply $2B + C = 50$. It says that, considering only flour suppies, the baker can choose to make 25 dozen muffins and no carrot cakes, or 50 carrot cakes and no muffins, or any one of innumerable other possibilities, just so long as the total amount of flour consumed is never more than 50 cups. Similar equations express other constraints, such as the limited supplies of honey and eggs.

With just two variables, all the equations can be drawn as straight lines on a two-dimensional graph *(see Figure 4.14)*. Each line divides the plane into two regions. One half encompasses the area where every point $(B, C)$ represents a possible answer; the other half is a forbidden zone. The complete collection of straight-line constraints fences in an area the shape of an irregular polygon within which the optimal answer must be found.

Which is the best of all possible answers? One way to find out is to compute for each pair of allowed values what the profit would be and then select the pair that gives the largest profit. But mathematics provides a convenient shortcut through this tedious procedure. In general, the optimal answer lies at one particular corner, or vertex, of the polygon. The coordinates of that vertex $(B, C)$ provide values of $B$ and $C$ for which the total profit will be highest.

Real-world business problems—from routing telephone calls to scheduling airline flights and allocating tasks to a work force—may involve hundreds or even thousands of variables and constraints. For these problems, the constraint equations form a complicated nest of hyperplanes in a high-dimensional space. As in the simpler two-variable case, each hyperplane cuts this enormous space in half. On one side of the surface are points corresponding to feasible plans of ac-

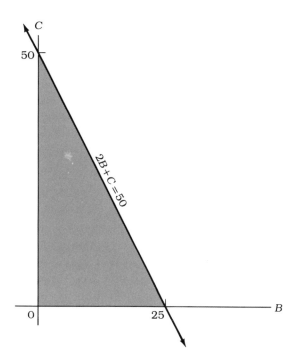

**FIGURE 4.14** Solving the bakery problem. Because $B$ and $C$ can't be negative, possible answers lie within the shaded area. Other constraints further restrict the area in which feasible values lie.

tion; on the other side lie infeasible points, which violate the constraint represented by the surface. Together, the intersecting hyperplanes create a multidimensional polyhedron called a *polytope*, which encloses all the feasible solutions. In this case, too, the best answer lies at a vertex.

Solving a linear programming problem is something like climbing to the top of an elaborate, irregular geodesic dome. Step by step, the appropriate mathematical algorithm steers the climber through a space that may be thousands of dimensions in extent to the peak of the polytope. What's needed is an efficient way to close in on the right corner.

The simplex method, introduced in 1947 by George B. Dantzig of Stanford University, navigates toward the answer by mathematically hopping from one vertex to an adjacent vertex across the polytope's surface, always taking the branch that leads "uphill," toward the "peak" vertex, which represents the optimum values for the given set of variables *(see Figure 4.15)*.

Using the simplex method effectively is virtually an art. Theoreti-

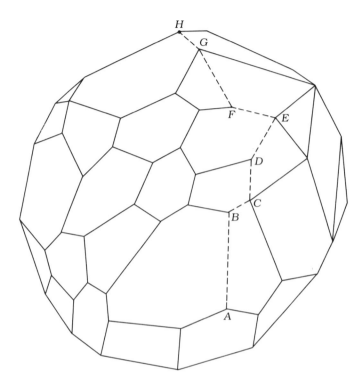

**FIGURE** 4.15    Starting at *A*, the simplex method proceeds from vertex to vertex until it reaches an optimal value at *H*.

cally, because the number of vertices to be visited may be immense, a computer could take practically forever to reach the right corner. In practice, the simplex method is surprisingly efficient, and years of work have refined it considerably. While just about any programmer can state the method's essence in a computer program of about two dozen lines, actual programs run to more than 50,000 lines. These elaborate steering mechanisms are replete with clever programming ploys that drive the algorithm to its goal more quickly. Only rarely, in a few pathological cases, is the method slowed down significantly because practically every vertex has to be visited before the peak vertex is at last attained.

Over the years, the simplex method has gradually been improved and has now reached the stage where one or another of its many variants can process problems with 15,000 to 20,000 constraints. Beyond this, the simplex algorithm generally becomes prohibitively slow and cumbersome. Unfortunately, many of today's business problems—especially those in the telecommunications industry—have far more than 20,000 constraints and variables.

In 1979, Russian mathematician L. G. Khachian introduced a method that seemed to overcome some of the simplex method's limitations. Khachian's method requires the construction of high-dimensional ellipsoids (in two dimensions, they would be ellipses) around the entire structure. These enveloping, high-dimensional balloons "slide" sideways under the influence of the polytope's constraining surfaces, so that the centers gradually close in on the optimal solution. This procedure guarantees that a computer will complete its job within what computer scientists call *polynomial* time, whereas the simplex method, at its worst, runs in *exponential* time. The same kind of time constraints are important for evaluating methods for factoring numbers.

The difference between polynomial time and exponential time is like the difference between $x^3$ and $3^x$. Say $x$ is 10; then $x^3$ is 1,000 and $3^x$ is 59,049. As the value of $x$ (for example, the number of variables in a linear programming problem) increases, exponential time quickly becomes astronomical. On worst-case problems for the simplex technique, when the problem size is doubled, the time required to process such a problem is squared. For the ellipsoid method, when the problem size doubles, the processing time for a worst-case problem goes up by a smaller factor.

However, the Russian algorithm proved disappointing: in practice, it turns out to be far slower than the simplex method for most everyday problems and has an advantage only with very large numbers of variables. Except in certain pathological cases, the simplex method completes the job in about as many steps as there are variables. The worst case, in which every vertex must be visited, rarely comes up. The ellipsoid method, however, shows no such convenient behavior.

In the last few years, a new, remarkably efficient method, discovered by Narendra K. Karmarkar of AT&T Bell Laboratories, has taken the spotlight. Karmarkar's algorithm boldly jumps away from the concept of a surface path zigzagging from vertex to vertex on a polytope of fixed shape. Instead, Karmarkar's algorithm plunges into the polytope's interior and works from the inside, guided by step-by-step transformations of the polytope's shape.

Karmarkar's search for the peak starts with a point that lies within the polytope. The first iteration twists the polytope, according to the rules of projective geometry, into a new shape centered around the starting point. From this newly established, central vantage point, the algorithm takes a step, following the path of steepest ascent or descent, to the next approximate solution point within the polytope's interior. That's done by inscribing a sphere within the polytope (like blowing up a balloon inside a box until it just fits without being scrunched) and selecting a new point near the sphere's surface in the direction of the best solution. Then the polytope is warped again to bring the new point to the center. Successive applications of these projective transformations have the effect of magnifying the parts of the feasible region close to the optimal vertex. In only a few steps, Karmarkar's method converges on the optimal solution.

Bounds, called *performance guarantees*, strictly limit how long an algorithm takes to do its job in the worst possible case and are important in theoretical computer science. Like the ellipsoid method, Narendra Karmarkar's scheme also runs in polynomial time, but the exponent governing how long his method takes is smaller than the exponent governing the ellipsoid method. The bound on Khachian's algorithm grows as $n^6$, rendering it impractical for large problems involving thousands of variables. Karmarkar's method grows as $n^{3.5}$, a significant improvement over the previous polynomial-time algorithm. As a result, when the number of variables in a problem increases, the running time for Karmarkar's method doesn't rise as rapidly as it does for the ellipsoid method. Moreover, the new method, when cleverly implemented in a computer program, in many cases seems to perform significantly faster than the simplex algorithm.

Speed does make a difference. The old joke that it takes 3 days of calculations to predict tomorrow's weather pinpoints the crucial problem. Using known methods and algorithms, many complex problems take too long to solve. Although the simplex method was, for a long time, the best technique known for solving linear programming problems, it had its limits. Karmarkar's method pushes up the number of variables and constraints that can be handled within a reasonable time. Such a vast increase in the number of variables that can be handled in a linear programming problem may reshape many theories in economics and perhaps other sciences. Problems that people once avoided because there was no method for solving them may be now within reach.

9 Culture Clash

11    Basins of Attraction

12    Bursting into Chaos

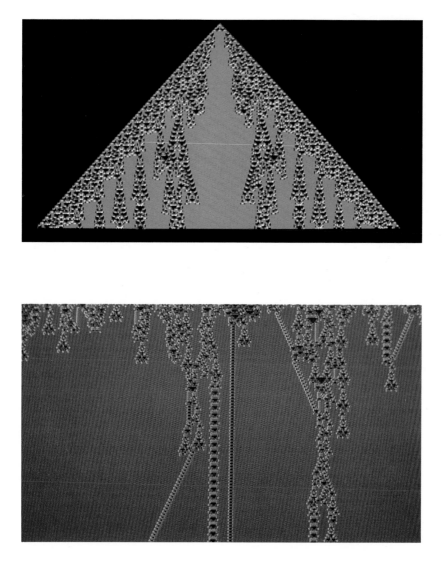

**13**    Grid Rules

**14    Snowflake Cells**

**15**   Probabalistic Drip

**16**    Crystal Tiles

# 5

## ANTS
## IN
## LABYRINTHS

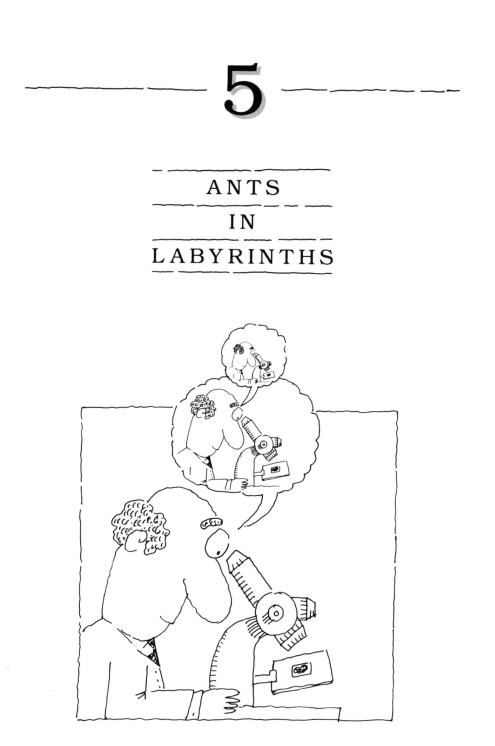

The flickering flames of a campfire highlight the jagged forms of nearby rocks. The smell of frying fish weaves through the still air. Clusters of gnarled pines crouch on the ragged, stony shore. Patches of golden wildflowers, scattered across a meadow, glow in the last light of a weary sun. At the far end of the lake, a saw-toothed range of towering mountain peaks, crowned with agglomerated ice, tear into the sky. From distant, billowy clouds, a flash of lightning zigzags through the air.

This mountain landscape, like many natural scenes, has a roughness that's hard to capture in the classical geometry of lines and planes, circles and spheres, triangles and cones. Euclidean geometry, created more than 2,000 years ago, best describes a human-made world of buildings and other structures based on straight lines and simple curves. Although smooth curves and regular shapes represent a powerful abstraction of reality, they can't fully describe the form of a cloud, a mountain, or a coastline. In the words of mathematician Benoit B. Mandelbrot, "Clouds are not spheres, mountains are not cones, coastlines are not circles, and bark is not smooth, nor does lightning travel in a straight line."

## —— — THE LANGUAGE OF NATURE —— ——

A close look shows that many natural forms, despite their irregular or tangled appearance, share a remarkable feature on which a new geometry can be hung. Clouds, mountains, and trees wear their irregularity in an unexpectedly orderly fashion. Nature is full of shapes that repeat themselves on different scales within the same object.

A fragment of rock looks like the mountain from which it was fractured. Clouds keep their distinctive appearance whether viewed from the ground or from an airplane window. A tree's twigs often have the same branching pattern seen at the tree's trunk. Elms, for instance, have two branches coming out of most forks. In a large tree, this repeated pattern, on ever smaller scales, may go through seven levels, from the trunk to the smallest twigs. Similar branching structures can be seen in the human body's system of veins and arteries, and in maps of river systems.

In all these examples, zooming in for a closer view doesn't smooth out the irregularities. Instead, the objects tend to show the same degree of roughness at different levels of magnification. Mandelbrot, the first person to recognize how extraordinarily widespread this type of structure is in nature, introduced the term self-similar to describe such objects and features. No matter how grainy, tangled, or wrinkled they are, the irregularities are still subject to strict rules.

In 1975, Mandelbrot coined the word *fractal* as a convenient label for irregular and fragmented self-similar shapes. Fractal objects contain structures nested within one another. Each smaller structure is a miniature, though not necessarily identical, version of the larger form. The mathematics of fractals mirrors this relation between patterns seen in the whole and patterns seen in parts of that whole.

Fractals turn out to have some surprising properties, especially in contrast to geometric shapes such as spheres, triangles, and lines. In the world of classical geometry, objects have a dimension expressed as a whole number. Spheres, cubes, and other solids are three-dimensional; squares, triangles, and other plane figures are two-dimensional; lines and curves are one-dimensional; and points are zero-dimensional. Measures of size — volume, area, and length — also reflect this fundamental classification.

Fractal curves can wriggle so much that they fall in the gap between two dimensions. They can have dimensions anywhere between one and two, depending on how much they meander. If the curve more closely resembles a line, then it is rather smooth and has a fractal dimension close to 1. A curve that zigzags wildly and comes close to filling the plane has a fractal dimension nearer to 2.

Similarly, a hilly fractal scene can lie somewhere between the second and third dimensions of classical geometry. A landscape with a fractal dimension close to 2 may show a huge hill with tiny projecting bumps, whereas one with a fractal dimension close to 3 would feature a rough surface with many medium-size hills and a few large ones. A higher fractional dimension means a greater degree of complexity and roughness. But a fractal dimension is never larger than the euclidean dimension of the space in which the fractal shape is embedded: a hilly scene would never have a dimension greater than 3. In general, fractal geometry fills in the spaces between whole-number dimensions.

Maps of a rugged coastline illustrate another curious property of fractals. Finer and finer scales reveal more and more detail and lead to longer and longer coastline lengths. On a world globe, the eastern coast of the United States looks like a fairly smooth line that stretches somewhere between 2,000 and 3,000 miles. The same coast drawn on an atlas page showing only the United States looks much more ragged. Adding in the lengths of capes and bays, its extent now seems more like 4,000 or 5,000 miles. Piecing together detailed navigational charts to create a giant coastal map reveals an incredibly complex curve that may be 10,000 or 12,000 miles long. A person walking along the coastline, staying within a step of the water's edge, would have to scramble more than 15,000 miles to complete the trip. A determined ant taking the same shoreline expedition but staying only an ant step away from the water may go 30,000 miles. Tinier coastline explorers even closer to the shoreline would have to travel even farther.

This curious result suggests that one consequence of self-similarity is that the simple notion of length no longer provides an adequate measure of size. Although it's reasonable to consider the width of a bookcase as a straight line and to assign to it a single value, a fractal coastline can't be considered in this way. Unlike the curves of euclidean geometry, which become straight lines when magnified, the fractal crinkles of coastlines, mountains, and clouds do not go away when observed closely. If a coastline's length is measured in smaller and smaller steps or with shorter and shorter measuring sticks, its length grows without bound. Because it wiggles so much, the true length of a fractal coast is infinite. Length, normally applied to one-dimensional objects such as curves, doesn't work for objects with a fractional dimension greater than 1.

Fractal geometry doesn't prove that euclidean geometry is wrong. It merely shows that classical geometry is limited in its ability to represent certain aspects of reality. Classical geometry is still a handy way to describe salt crystals, which are cubic, or planets, which are roughly spherical and travel around the sun in elliptic orbits. Fractal geometry, on the other hand, introduces a set of abstract forms that can be used to represent a wide range of irregular objects. It provides mathematicians and scientists with a new kind of meter stick for measuring and exploring nature.

## —— TAMING MATHEMATICAL MONSTERS ——

In 1904, Swedish mathematician Helge von Koch created a mathematically intriguing but disturbing curve. It zigzags so much that a traveler set down anywhere along the curve's path would have no idea in which direction to turn. Like many figures now known to be fractals, von Koch's curve can be generated by a step-by-step procedure that takes a simple initial figure and turns it into an increasingly crinkly form.

The Koch, or "snowflake," curve starts innocently enough as the outside edge of a large equilateral triangle *(see Figure 5.1)*. The addition of an equilateral triangle one-third the size of the original to the middle of each side of the large triangle turns the figure into a six-

**FIGURE 5.1**    The first four stages in constructing a Koch snowflake.

pointed star. The star's boundary has 12 segments, and the length of its outer edge is four-thirds that of the original triangle's perimeter. In the next stage, 12 smaller triangles are added to the middle of each side of the star. Continuing the process by endlessly adding smaller and smaller triangles to every new side on ever finer scales produces the Koch snowflake. Any portion of the Koch snowflake magnified in scale by a factor of three will look exactly like the original shape.

The boundary of the Koch snowflake, so convoluted that it's impossible to admire in all its fine detail, is continuous but certainly not smooth. It has an infinite number of zigzags between any two points on the curve. A tangent — the unique straight line that touches a curve at only one point — can never be drawn anywhere along its perimeter. Moreover, the length between any two points is infinite, yet the curve bounds a finite area not much bigger than the area of the original equilateral triangle. In fact, despite the figure's wrinkled border, it's possible to compute that the area bounded by the curve is exactly eight-fifths that of the initial triangle.

Such strange mathematical behavior led mathematicians at the turn of the century to label this and several similar curves as mathematical monstrosities. But these monster curves are remarkably easy to generate. Pointing triangles inward rather than outward — subtracting instead of adding them at each step — produces the antisnowflake curve. This lacy form, too, has an infinitely long outer edge that intersects itself infinitely often, but its area is only two-fifths that of the starting triangle.

The same idea of adding or subtracting pieces that are successively smaller in size works for a square or any other polygon. It also works in three dimensions. Dividing each face of a regular tetrahedron into four equilateral triangles and erecting a smaller tetrahedron on each face's middle triangle, then continuing this step-by-step procedure indefinitely creates a prickly, three-dimensional analog of the Koch snowflake. Its surface area is infinite, but the figure bounds a finite volume. The supply of monsters seems limitless!

No wonder mathematicians were disturbed when they first encountered such bizarre behavior. These pathological curves and surfaces, they believed, were aberrations—skeletons in the closet of otherwise orderly mathematics. To them, such figments of the imagination represented a mathematical pathology having nothing to do with any possible real-world phenomenon and were unlike anything found in nature.

Nevertheless, a few mathematicians took these monster shapes seriously enough to explore their properties in some detail. In 1919, the German mathematician Felix Hausdorff suggested a way to generalize the notion of dimension, which put these disturbing forms into a class of their own. He came up with the idea of fractional dimension, a concept that is now one of several methods used to characterize a fractal.

The idea of fractal dimension extends the concept of dimension normally used for describing ordinary, regular objects such as squares and cubes. The idea is to figure out how many small objects or units of size $p$ are needed to cover a large object of size $P$. In the case of a line segment, say, 8 meters long, it takes eight 1-meter lengths to cover the whole line. If the measuring unit were 10 centimeters long, then it would take 80 such units to cover the 8-meter length. The ratio between the two results, 80 and 8, is 10:1 or $10^1$, which is the ratio of the measuring-stick lengths: 1 meter and 10 centimeters. The exponent 1 matches the dimension of a line. Similarly, in the case of area, a square 1 meter by 1 meter fits into an 8-square-meter area eight times, whereas a measuring square 10 centimeters by 10 centimeters fits into the same area 800 times. The ratio between the two results, 800 and 8, is 100, or $10^2$. The ratio of the measuring-square unit areas ($1^2$ m:$10^2$ cm) is also $10^2$, and the exponent 2 matches the dimension normally associated with area. Volume can be handled in the same way, using measuring cubes of different sizes. In every case, the dimension of the object appears as the exponent of the ratio of the length scale of the measuring units.

The whole process of finding the dimension of an object can be turned into a mathematical operation of taking logarithms of the appropriate ratio. Tripling the width of a square creates a new square that contains nine of the original squares. Its dimension is calculated by taking logarithms of the size ratio, or magnification: log 9/log 3 = log $3^2$/log 3 = 2 log 3/log 3 = 2. Thus a square is two-dimensional. Doubling the size of a cube produces a new cube that contains eight of the original cubes. Its dimension is log 8/log 2 = log $2^3$/log 2 = 3 log 2/log 2 = 3. It's no surprise that a cube is three-dimensional.

In general, for any fractal object of size $P$, constructed of smaller units of size $p$, the number, $N$, of units that fits into the object is the size ratio raised to a power, and that exponent, $d$, is called the Hausdorff dimension. In mathematical terms, this can be written as

$$N = \left(\frac{P}{p}\right)^d \text{ or } d = \log N/\log (P/p).$$

This way of defining dimension shows that familiar objects, such as the line, square, and cube are also fractals, although mathematically they count as "trivial" cases. The line contains within itself little line segments, the square contains little squares, and the cube little cubes.

Applying the concept of the Hausdorff dimension to the Koch curve gives a fractional dimension. Suppose at the first stage of its construction, the snowflake curve is 1 centimeter on a side. With a resolution of 1 centimeter, the curve is seen as a triangle made up of 3 line segments. Finer wrinkles aren't visible. If the resolution is improved to $\frac{1}{3}$ centimeter, 12 segments, each $\frac{1}{3}$ centimeter in length, become evident. Every time the unit of measurement is cut by a factor of one-third, the number of visible segments increases four times. In this case, $N = 4$ and $P/p = 3$. Hence, $3^d = 4$, so $\log 3^d = \log 4$, $d \log 3 = \log 4$, and $d = \log 4/\log 3$. The Hausdorff dimension of the fractal Koch curve is 1.2618. . . . The strange properties of the snowflake curve stem from the fact that it is not a one-dimensional object. It belongs in the wonderland of fractional dimensions.

Fractals in nature typically lack the regularity evident in a Koch curve, but natural fractals are often self-similar in a statistical sense. With a large enough collection of examples, a magnified portion of one sample will closely match some other sample in the collection. The fractal dimension of these shapes can be determined only by taking the average of the fractal dimensions at many different length scales. Although the Koch curve's peninsulas and bays occur with absolute precision, they bear a striking resemblance to the shape of a rugged coastline. Although the bends and wiggles of a coastline are not in precisely the same locations on all scales, the coastline's general shape looks the same no matter what scale is used for measurement.

Another significant difference between a Koch curve and an actual coastline is that the curve is an idealized mathematical form with structure at infinitely many levels. A physical coastline isn't really an infinitely complex line. However, for many purposes, a fractal is a better, though not necessarily perfect, model for a coastline than is a smoother, finitely complex curve. The similarity between an abstract object like the Koch curve and the typical characteristics of a coastline is enough to define an approximate fractal dimension for actual ocean coastlines. Studying maps on different scales shows that ocean coastlines vary in dimension but generally range from 1.15 to 1.25 — not quite as rough as a Koch curve.

One of the joys of fractal geometry is the opportunity to create new monster curves and other pathological forms. Usually, that means starting with some basic shape or figure (the initiator), then applying a rule (the generator) that step-by-step makes the figure more irregular, tangled, or wrinkled on ever smaller scales in an endlessly looping process.

One strange form, called the Sierpiński gasket, starts as a triangle, like the Koch snowflake, but all the drawing and cutting take place within the figure. Marking the midpoints of the three sides and joining those points creates a new triangle embedded within the original triangle. This construction breaks up the large triangle into four smaller triangular pieces: one central and three corner triangles. Cutting out the central triangle leaves the three corner triangles. Then the process of finding midpoints and drawing in a triangle is repeated within each of the remaining triangles. The central triangle that sits within each of these corner pieces is cut out, leaving nine small triangles. The pattern seen in the first step is thus duplicated in each of the corner triangles within the figure. The process continues indefinitely to generate an arrangement of triangles nested within triangles nested within triangles, and so on, resulting in a two-dimensional sieve punctured by an infinite number of holes (*see Figure 5.2*).

At each step in the process, the length of a triangle's side is cut in half, and three times as many triangles appear. Therefore, the Sierpiński gasket has a Hausdorff dimension of $\log 3/\log 2 = 1.584$. . . . It also happens to have zero area.

A similar procedure can be applied to a square. Dividing each side of a square into three parts to create a 3 by 3 grid and removing the central square is the first stage in generating an infinitely moth-eaten Sierpiński carpet (*see Figure 5.3*). Its fractal dimension is $\log 8/\log 3 = 1.8928$. . ., showing that this figure is actually more like a bumpy curve than a real carpet.

The same operation can take place in three dimensions. Starting with a cube, dividing it into 27 smaller cubes, and removing the central cube as well as the cubes lying at the center of each face of the

**FIGURE 5.2** This fractal starts as a triangle. Removing successively smaller triangles step by step leads to the remarkable figure known as the Sierpińksi gasket, which turns out to have zero area.

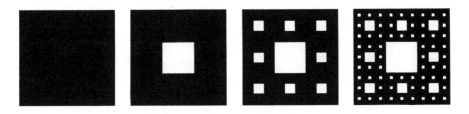

**FIGURE 5.3** A Sierpiński carpet begins as a square. Removing square patches generates the fractal form.

original cube (7 in all) leads inexorably, step by step, to a novel form called the Menger sponge *(see Figure 5.4)*. Its dimension is log 20/log 3 = 2.727. . . . Because its dimension is closer to three than to two, the Menger sponge is more a solid body than a smooth surface.

At the opposite dimensional extreme, a similar removal process can be applied to a one-dimensional line segment. In this case, a section is removed from the middle of a line. Then a corresponding

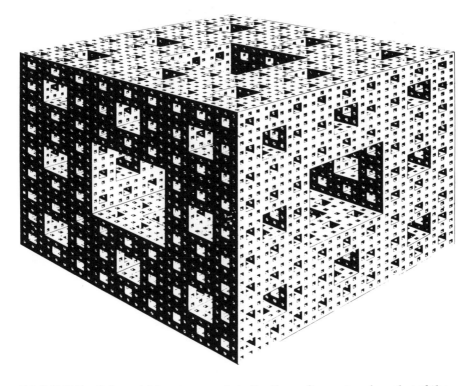

**FIGURE 5.4** A Menger sponge is the three-dimensional analog of the Sierpiński carpet.

section is removed from the middle of the two remaining pieces, and so on, until the line falls apart in a shower of dimensionless fragments. This extremely simple fractal is an example of a Cantor set, named for mathematician Georg Cantor. All Cantor sets created by breaking up a line have a dimension between zero and one.

Mandelbrot, who initiated much of the research on fractals, has been particularly inventive in generating novel fractals with unique properties. Some of these fractals resemble natural shapes, such as mountains. Others have geometric features that make them useful as idealized models for physical processes, such as the seeping of oil through porous rock or the way in which a material becomes magnetized. Until Mandelbrot recognized fractal forms as a potentially rewarding tool for analyzing a variety of physical phenomena, these forms remained bizarre and useless curiosities.

The first application of fractals was to solve the problem of noise during data transmission. Like the irregular crackling sound sometimes heard during a radio broadcast, electrical disturbances interrupt and confuse the flow of data over telephone lines and other transmission channels. By looking at the pattern of errors that occur when computer data are transmitted electrically as a group of on/off signals, Mandelbrot noticed that the errors seemed to show up in bursts. Examining these bursts more closely, he found that each burst was itself intermittent. Those shorter bursts and their associated gaps also had a similar structure. The best mathematical model for these characteristic bursts appeared to be the Cantor set.

The Cantor set can be extended to three dimensions to become what Mandelbrot calls a Cantor dust. This fractal dust starts off as a solid block of matter, which is divided into stacks of smaller blocks. Some of the smaller blocks are randomly removed. The remaining blocks are subdivided further, and even more matter is randomly removed. Gradually, this Swiss-cheese structure comes to resemble water droplets scattered in a cloud or even the clusters of stars and galaxies dispersed throughout space. Ultimately, it becomes a dust of totally disconnected points.

Using the Cantor-dust concept, Mandelbrot concocted distributions for stars in a galaxy and galaxies in the universe. Although the model is a forgery and uses no real astronomical data, it closely resembles the hitherto inexplicable distributions of stellar matter seen by astronomers. Astrophysicists have now confirmed that as the model predicts, mass within the universe is distributed throughout space like a three-dimensional Cantor set, with large regions of space left empty. Cantor-set fractals describe not only the way matter clusters in space but also the way it clusters in time. Cantor sets seem to describe cars on a crowded highway, cotton price fluctuations since the nineteenth century, and the rising and falling of the River Nile over more than 2,000 years.

As the search for fractals branches into ever more exotic trails, the concept of fractal is also shifting. Mathematicians are finding that the idea of self-similarity alone doesn't cover all possible fractals. In 1980, Mandelbrot himself described a fractal set that John Hubbard later named the Mandelbrot set. It has the appearance of a snowman with a bad case of warts and comes up when simple mathematical expressions such as $x^2 - 3x$ are repeatedly evaluated, starting with an initial value for $x$, then substituting the resulting answer back into the original equation, and so on. On superficial inspection, the Mandelbrot set looks like a self-similar fractal, with infinitely many copies of itself within itself. On detailed investigation, however, the set is extraordinarily complicated. The baby Mandelbrot sets within the parent Mandelbrot set are fuzzier than the original. They have more "hair" and other curious features. Their own baby Mandelbrot sets are fuzzier still. This hairiness increases at finer and finer scales until it seems to carpet whole areas.

Fractals such as the Mandelbrot set are called nonlinear fractals. For self-similar fractals, lines that show up within a figure, whether blown up or reduced in size, remain lines. For nonlinear fractals, such a change in scale doesn't necessarily preserve the straightness of individual lines. Instead, changes in scale reveal a host of fascinating new features, many of which make an appearance in the next chapter.

Mathematicians have come across a broad range of objects that have the property of infinite irregularity characteristic of fractals. Such forms ultimately raise a not yet satisfactorily resolved theoretical question: What should or should not be included within the compass of the term "fractal"?

## ———— —— PACKING IT IN —— —— ————

Fractal geometry and computer graphics are inextricably linked. Computer graphics provides a convenient way of picturing and exploring fractal objects, and fractal geometry is a useful tool for creating computer images. The simple, repeated operations that go into the construction of a fractal are ideally suited to the way a computer functions. The computer patiently performs the same set of operations over and over again to generate a particular fractal object. Of course, the fractal image that appears on a computer's video display isn't really, in a mathematical sense, a true fractal. A computer generating a Sierpiński gasket, for instance, can display no triangles smaller than the tiny spots of light, or pixels, that make up the screen. The true Sierpiński gasket is filled with even smaller triangles. Computer graphics therefore provides a useful but only approximate picture of fractal forms.

Fractal geometry also provides an elegant, efficient way to draw realistic natural objects on a video screen. It overcomes some of the disadvantages that crop up when more conventional geometric techniques for creating a video image are used. One common conventional method is to piece the image together from simple euclidean shapes such as squares, triangles, and circles. Such constructions work well for angular, human-made objects such as bridges, robots, and spacecraft, but they can't capture the enormous detail and irregularity evident in clouds or natural terrain.

Specifying all that detail is a monumental computing task requiring either a lengthy computer program or one that would need huge amounts of time to compute a given scene. Fractals suggest a simpler solution. The fractal approach to drawing pictures involves feeding into a computer a small set of numbers that generates a basic shape, such as a triangle. That shape is then recreated many times on smaller and smaller scales within the original figure. Random variations are thrown in to make the image look a little rougher and, as a result, more realistic. Such artificial landscapes can be mathematically magnified to reveal more detail, just as a close-up lens probes deeply into a natural scene.

One way to build a mountain using loosely based fractal geometry concepts is to start with a triangle. The computer first finds the midpoint of each of the triangle's sides. Each midpoint is then displaced along its corresponding edge through a distance determined by a random-number generator. Joining the displaced midpoints generates a new triangle and divides the original figure into four smaller triangles. Unlike the triangles in the Sierpiński gasket, these component triangles aren't necessarily identical or even equilateral. The same procedure is applied in turn to each of the four new triangles, generating sixteen triangles, to each of which the procedure is applied again, and so on. The process continues until the individual triangles are so small that their edges can no longer be distinguished. Although the algorithm for subdividing the triangles is straightforward, the resulting figure is a complex polygonal surface. The addition of color and shadow turns this figure into a reasonable facsimile of a mountain (see Figure 5.5).

To generate a fractal landscape, the designer can enter into the computer a set of elevations, specifying the locations of mountain peaks and valleys. The computer then connects these points to pro-

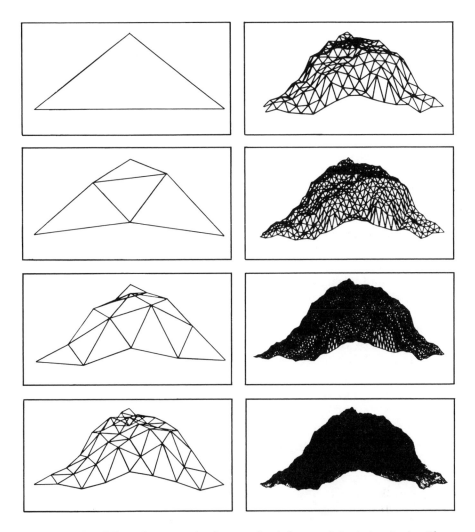

**FIGURE 5.5** One way to draw a fractal mountain is to start with a network of triangles and then break that down into finer nets to generate an irregular surface.

duce a complicated polygon. That figure is further subdivided into a mesh of simple triangles. Now the computer subdivides each triangle into smaller triangles, using the same technique as for drawing a single mountain. In the end, the computer screen shows an irregular terrain made up of a large number of triangular facets. Given sufficient computing time, the facets can be subdivided to the point where their edges are too small to be distinguished.

This particular recipe for building a mountain landscape is one of the simplest of several possible construction schemes. Mandelbrot's own scheme, which sticks more closely to the true characteristics of a fractal, generates somewhat more realistic scenes — ones that wouldn't be mistaken for a mound of crumpled paper. (The price of this greater realism, however, is longer computation time to generate the image.)

Mandelbrot and his colleagues start with a mathematical construct that closely resembles a *random walk* — a sequence of up and down steps, determined by the toss of a coin. A set of these jagged, vertical cross sections is then assembled to form a markedly rough landscape. A final mathematical smoothing operation gives the constructed scene a more natural look. Any points on a random-walk cross section that happen to go below an arbitrarily set "zero" level are automatically reset to zero. Collections of these points appear as depressions filled with water *(see Figure 5.6 and Color Plate 5).*

Clouds can be created by putting together pictures of selected components of white noise, which corresponds to the hiss heard between FM radio stations. White noise, which can be described by fractal geometry, consists of fluctuations spread evenly throughout the radio-frequency spectrum. Selecting a fractal dimension of 3.2 puts together a combination of long-wave and short-wave components that, when plotted, produce the proper puffiness for a cloud. With the addition of the correct lighting and coloring, the result is a soft-looking, wispy object.

**FIGURE 5.6**    A fractal mountain like the one shown can be constructed from the contours generated by a random walk.

Fractal coastlines can be generated in much the same way that clouds are created. Both clouds and coastlines have the same basic, computer-generated relief or outline. The key difference lies in the type of lighting applied to the fractal outline. Light reflects directly off the surface of a landscape, but it partially penetrates a cloud and scatters, or softly diffuses. A computer program can be written to mimic these different lighting effects.

Snowflakes can be made by using a fractal branching program to generate a tree structure. The treelike form is then reproduced six times, in the same way that a kaleidoscope creates a picture with sixfold symmetry. By changing certain parameters in the branching routine, a wide variety of different snowflake designs can be created *(see Figure 5.7)*. However, because a snowflake isn't really a true fractal, the images don't always look quite right.

Attempts to generate realistic fractal models of trees have had

**FIGURE 5.7**   Generating a fractal pattern, then reproducing it six times, results in a snowflakelike form.

only limited success so far. As trees grow, twigs and branches tend to avoid one another and to die off when severely overshadowed. The combination of randomness and a strong degree of self-interaction greatly complicates the generation of acceptable fractal images of trees.

The fractal images just described have been created largely by trial and error and are the result of computer doodlers' stumbling accidentally on mathematical procedures, or algorithms, that happen to lead to drawings that look like natural objects. The trial-and-error period, however, is reaching an end as more systematic approaches to creating fractal images are developed.

Michael Barnsley of the Georgia Institute of Technology has tackled the problem of finding a specific fractal to fit a given natural object. He and his coworkers have been studying how a scene's geometry can be analyzed to generate an appropriate set of rules that can then be used to recreate the scene. Because fractal mathematics is a compact way to store the characteristics of an object, this approach would compress the content of an image into just a few equations.

The idea is to start with a digitized picture. Such a picture may consist of a 1,000 by 1,000 grid of dots or pixels. Each pixel is assigned, say, eight bits of data to represent 256 different shades of gray or an equal number of different colors. Thus, the entire picture can be thought of as a string of 8 million ones and zeros, one digit for each bit. If this string of digits can be encoded in some way to produce a new, shorter string of digits, then the image is said to have been compressed. Of course, the compressed string should be able to reproduce, pixel for pixel, the original picture.

Barnsley and his colleagues believe that the key to image compression is in the redundancy found in natural forms. For instance, one pine needle is more or less like any other pine needle. That means that there's really no need to describe each needle over and over again. It's sufficient to describe just one. Fractals, in which similar structures are repeated at smaller scales within an object, also capture much of the redundancy found in nature.

Several important mathematical concepts lie at the heart of Barnsley's scheme. His procedure relies on mathematical operations called affine transformations. An *affine transformation* behaves somewhat like a drafting machine that takes in a drawing—that is, takes in the coordinates for all the points making up the lines in the drawing—then shrinks, enlarges, shifts, rotates, or skews the picture and finally spews out a distorted version of the original.

Affine transformations can be applied to any type of object, including triangles, leaves, mountains, ferns, chimneys, clouds, or even the space in which an object may sit. In the case of a leaf, the idea is to find smaller, distorted copies of the leaf that, when fitted together and piled one on top of another so that they partially overlap, form a "collage," which approximately adds up to the original, full leaf. Each distorted, shrunken copy is defined by a particular affine transformation, or "contractive map," as it's called, of the whole leaf. If it takes four miniature copies of the leaf to approximate the whole leaf, then there will be four such transformations *(see Figure 5.8, top)*.

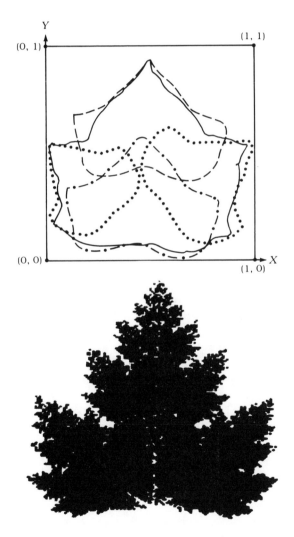

**FIGURE** 5.8  *Top,* approximate collage of a leaf with smaller distorted copies of itself; *bottom,* an image of a leaf generated by randomly applying a set of affine transformations.

Now the original image or "target," whether leaf or cloud, can be thrown away, leaving only the corresponding collection of affine transformations. These can be used to recreate the original image, essentially by molding a piece of space. That's done by starting with a point somewhere on a computer screen and applying one of the available transformations to shift the point to a new spot. That spot is marked. Again, randomly applying one of the transformations shifts the point to another location. The new spot is colored in, and the process is repeated again and again.

Amazingly, although the designated point appears to hop about aimlessly, first pulled one way then another, a pattern gradually emerges. The point's colored tracks add up to an image called an *attractor*—a concept discussed in more detail in Chapter 6. In the case of the four leaf transformations, the attractor is an object that looks very much like the original leaf *(see Figure 5.8, bottom)*.

How this part of the process works can best be understood using a simple example on a sheet of squared graph paper. Imagine a rectangle with three of its corners labeled 1, 2, and 3. Suppose there are also three transformations. The first shrinks everything toward corner 1, the second shrinks everything toward corner 2, while the third shrinks everything toward corner 3, always by a factor of one-half. Now, select a starting point somewhere on the grid. Randomly apply one of the three transformations. That transformation will designate a new point halfway between the original point and one of the corners, depending on the choice of transformation. Randomly applying a second transformation (it may be any one of the three available) locates another point halfway between the previous point and the appropriate corner. As you chase the point around the sheet of graph paper, marking each landing spot, you are producing a somewhat distorted Sierpiński gasket.

It turns out that any particular collection of affine transformations, when iterated randomly, produces a unique fractal figure. The trick is to find the right group of tranformations to use for generating a particular image. That's done using the "collage" process, for example, a leaf covered by little copies of itself. Furthermore, the probability of using a certain tranformation need not be the same as the probability of applying any other transformation in the set. And because some grid squares are likely to be visited more often than others, keeping track of the relative number of visits to each square provides a way to specify color brightness and intensity or to define a gray scale. In this way, a lot of information is packed into a few formulas, and you can compress images by encoding them as a collection of rules. Randomly iterating the rules then recreates the image.

Using his technique, Barnsley and his group have been able to create remarkable three-dimensional renderings of natural objects such as ferns. Their graceful model of a black spleenwort fern *(see Figure 5.9 and Color Plate 6)* is the product of the application of a collage of four affine transformations — each a combination of a translation, a rotation, and a contraction. When applied to a given point $(x, y)$, a particular transformation generates a new point: $[rx(\cos A) - sy(\sin B) + h, rx(\sin A) + sy(\cos B) + k]$. The parameters $h$, $k$, $A$, $B$, $r$, $s$ have the values shown in the following table:

| Map | Translations | | Rotations | | Scalings | | Probabilities |
| | h | k | A | B | r | s | |
|---|---|---|---|---|---|---|---|
| 1 | 0.0 | 0.0 | 0 | 0 | 0.0 | 0.16 | .005 |
| 2 | 0.0 | 1.6 | −2.5 | −2.5 | 0.85 | 0.85 | .8 |
| 3 | 0.0 | 1.6 | 49 | 49 | 0.3 | 0.34 | .0975 |
| 4 | 0.0 | 0.44 | 120 | −50 | 0.3 | 0.37 | .0975 |

Barnsley has also been able to come up with reasonable fractal reproductions of photographs. In one instance, he uses 57 affine transformations, or maps, as they are often called, and four colors — a total of 2,000 bytes of information — to model three chimneys set in a landscape against a cloudy sky. "The idea is that we can fly into this picture," Barnsley says. "You can pan across the image, you can zoom into it, and you can make predictions about what's hidden in the picture."

As you blow up the picture to show more and more detail, parts of the picture degenerate into nonsense, but some features, such as the chimneys, the smoke, and the horizon, remain reasonably realistic, even when the image compression ratio is more than 10,000 to 1. The degree of compression attained depends on how much of the regenerated picture makes sense.

Interestingly, the images created using this scheme aren't self-similar. Instead, they're self-affine because an object such as one of Barnsley's fractal ferns shows slightly different features on different scales. Magnifying the image reveals subtle differences in form and color, and the magnification can be carried on indefinitely, as for any fractal.

If image compression can be made as effective as Barnsley's work suggests and if the whole process can be automated, it could provide an efficient means for storing data in a computer's memory, for transmitting photographs over telephone lines, for recognizing specific ob-

**FIGURE** 5.9    A set of simple affine transformations instructs a computer to generate this image of a black spleenwort fern.

jects in a landscape, and for simulating natural scenery on a computer. Someday, it may even be possible to convey a movie from one computer to another simply by sending a chain of formulas down a telephone line.

## FRACTAL EXCURSIONS

When the delicate fragrance of a perfume weaves through the air, individual perfume molecules, jostled about haphazardly in the hurly-burly of their collisions, follow a jagged path. Air currents further tangle these tortuous paths into monstrous trails. This type of irregular motion can also be seen in the unceasing, restless dance of tiny, barely visible particles suspended in a liquid. Such jitteriness is known as brownian motion.

Theoretically, a scientist can trace a brownian particle's irregular trajectory by periodically plotting the particle's precise location. The result is a string of points strewn across a sheet of graph paper *(see Figure 5.10)*. But using straight lines to connect the picture's dots fails to capture the motion's true intricacy. Between every pair of plotted points lies another jagged path. That's just the kind of motion whose essential features can be compactly expressed by fractal geometry.

This type of motion is akin to a random walk—a sequence of steps whose size and direction, as we saw earlier, are determined by chance. In a simple, one-dimensional version of a random walk, a person flips a coin and takes one step forward if the result is heads and one step backward if the result is tails. Across a broader stage, a random walk looks even more like the steps of a drunken sailor.

When a brownian trajectory confined to a plane is examined

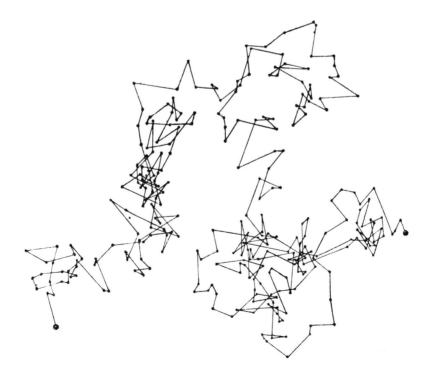

**FIGURE** 5.10    A particle exhibiting brownian motion zigzags randomly along a path that resembles a fractal.

increasingly closely, its length, like that of a coastline, grows without bound. In fact, the trail itself ends up filling practically the whole plane in which the motion takes place. Although the trajectory itself is a one-dimensional curve, the path's tendency to fill the plane marks it as a fractal of dimension 2.

Fractal concepts can be used to describe not only an astonishing array of fragmented or branching natural structures but also the dynamic properties of these structures—from the movement of Brownian particles to the drip of scalding water through coffee grounds. In a way, fractals represent a new kind of meter stick that scientists can use to measure natural phenomena. For example, they can use fractals to study the way materials are put together, the way they shatter, the way they branch, and the way they conduct heat or electricity.

A word of caution is warranted, however. Mathematical fractals have properties that aren't actually found in natural objects. No real structure can be magnified repeatedly an infinite number of times and still look roughly the same. One reason is the finite size of atoms and molecules. Another is that real objects may, at some magnification, abruptly shift from one type of structural pattern to another. Nevertheless, fractal models provide a useful approximation of reality, at least over a finite range of scales.

The microscopically jagged surface of a piece of fractured metal is one example of a material property that lends itself to fractal analysis. A few years ago, Benoit Mandelbrot worked with some metallurgists to come up with a method for specifying the roughness of a given surface. They found that a large number of broken metal surfaces, although not all, have a roughness that can be represented by a fractal dimension. Experiments showed that the measured fractal dimension takes on the same value for different specimens of identically treated samples of the same metal. The investigators also discovered that different heat treatments not only affect the toughness of a metal but also change its fractal dimension. They concluded that a metal surface's fractal dimension may itself be a useful measurement of a metal's toughness or strength, providing metallurgists with a new tool for characterizing metals.

Fractal ideas have also encouraged scientists to look anew at old, seemingly inexplicable experimental results once destined for the wastebasket and to reexamine problems that previously looked so complicated that they were ignored. Physicists and other researchers now realize that many formerly puzzling results actually reflect the dimensions of fractal geometric objects. With this insight, some apparently complex problems become relatively simple.

In analyzing experimental data, scientists usually look for simple relationships between variables; such as the relationship between the intensity of sound waves scattered from a metal surface and the waves' frequency. If a theory predicts that doubling the frequency will quadruple the intensity, the intensity should be proportional to the frequency squared. However, in many experiments, the exponents that express the proportionality turn out to be numbers like 2.79 instead of integers. Scientists who were taught to think of integers as the natural way to represent physical processes are only now beginning to see that noninteger exponents are just as likely, if not more likely, to turn up in nature.

An example from the world of electrochemistry shows the interaction between experimental observation and a mathematical model. For decades, scientists had noticed that the interface between a metal electrode and the electrically conducting liquid, or electrolyte, in which the electrode is immersed has strange electrical properties. Simple electrical theory predicts that the interface should behave like a capacitor, an electrical device that stores then abruptly releases electrical charge. When an alternating current passes between electrode and electrolyte, it ought to meet a resistance that is simply related to the frequency. However, experimental studies show that this resistance is actually inversely proportional to the frequency raised to a fractional power between 0 and 1.

Further study shows that the resistance appears to depend on the roughness of the electrode surface. As the surface is made smoother, the exponent approaches 1. But, under magnification, even well-polished electrode surfaces still show long grooves with jagged cross sections and edges. The grooves themselves have finer scratches, suggesting a degree of self-similarity.

The theorist's job is to come up with a suitable mathematical model that captures an electrode's observed characteristics and its consequent behavior. That mathematical model must be simple and regular enough to be mathematically solvable but not so far removed from an electrode's observed features that any mathematical results from applying the model would be suspect. The idea is not to paint a realistic portrait but to capture the spirit of the phenomenon with a caricature. In that way, researchers may catch a glimpse of what is actually happening at the interface.

Given the nature of the observed electrode grooves, a good place to start is with a fractal. But which fractal? There are many to choose from. Some are easier to handle mathematically than others. One good fractal candidate, related to the Cantor set described earlier, is called the Cantor bar and has a dimension less than 1.

The Cantor bar begins as a thick, solid line segment. Taking a chunk out of the center of the bar breaks it into two pieces. The two resulting fragments are in turn broken according to the same rule, and the process is repeated ad infinitum *(see Figure 5.11, left)*. If the length of each broken piece is $\frac{1}{a}$ times the size of the original piece, where $a$ is greater than 2, then the object has a fractal dimension of log 2/log $a$, which must be less than 1.

When the increasingly fragmented bars are welded together — starting with the full, original bar on the bottom, then adding the next two pieces, then topping those with the next four broken pieces, and so on — the result is a kind of symmetric urban landscape with stepped towers stretching higher and higher as they get thinner and thinner *(see Figure 5.11, right)*. Although the surface of an electrode isn't nearly this regular, the mathematical model captures some of the features of a rough, grooved surface, especially the idea of grooves within grooves within grooves.

Giving electrical properties to this geometric pattern allows the calculation of various quantities that can also be measured experimentally. Currents, for instance, can pass directly across the interface or out along the grooves. That leads to a simple equation predicting that the interface resistance is inversely proportional to the frequency raised to a certain power, and the exponent depends on the geometry of the surface. Now, it's up to experimentalists to see how well this fractal theory matches what's seen in the laboratory. The theorist would then be able to refine the model further and to suggest which measurements would more likely reveal the processes occurring at the electrode-electrolyte interface.

Fractal models also play a useful role in percolation processes. The term conjures up an image of brewing coffee or the trickle of a liquid through a gravel bed, but it also applies to an important class of structures known as percolation clusters. Such clusters have properties similar to those of a floor covered with a random mixture of copper and vinyl tiles. Current will flow from one side of the floor to the other if there is a continuous copper path, no matter how round-

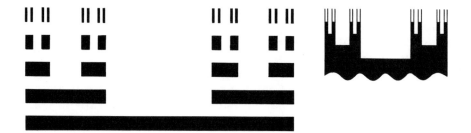

**FIGURE 5.11** *Left,* the first five members of a Cantor bar; *right,* a Cantor-bar model for a rough surface.

about. If most of the tiles are vinyl, current isn't likely to go through. Adding more randomly placed copper tiles increases the likelihood that patches of copper tiles will be linked to form a continuous path. A percolating cluster represents the point at which a conducting path is first created. Percolation clusters are particularly useful for modeling processes such as the seeping of oil through porous rock and the spread of plant species in a meadow.

This construct seems to work as a mathematical metaphor for many materials in which different substances having antagonistic properties are intermingled. Such materials may contain random mixtures of electrical conductors and insulators, magnetic and non-magnetic components, or elastic and brittle parts. The critical concentration of the two materials in the cluster, not the simple average of the properties of the components, determines which property predominates and how the mixture behaves. That in turn decides physical characteristics, such as the structure of some polymers, the conductivity of alloys, the efficiency of telephone networks, the propagation of forest fires and infectious diseases, and the spread of plant species in an ecosystem. Percolation models also provide useful pictures of abrupt changes in phase, such as the lining up of atomic spins to create a magnet, or the sudden, temperature-dependent onset of superconductivity in a thin metallic film.

Not only are the results of calculations done on an actual, random percolation cluster hard to understand, but the computations use up a great deal of computer time and are only approximate because of the complexity of the process. Researchers therefore look for fractal patterns that are systematic enough to ease mathematical calculations but random enough to be realistic.

Fractals also come up in percolation problems in another way. In a problem first proposed in the sixteenth century, an orchard is laid out in the shape of a checkerboard, with each space occupied by a tree. The spaces are small enough that neighboring trees touch each other. Tragically, a plague invades the orchard and spreads from tree to tree, destroying the entire fruit crop. The question at the time of replanting is how many trees should be excluded from the grid to prevent a disease from spreading from one end of the orchard to the other, while still maximizing the fruit yield.

Over the years, various attempts to answer the question have led to a best estimate that about 41 percent of the checkerboard squares should remain unoccupied. The critical concentration for percolation (when the disease can make its way from tree to tree across the orchard) is 59.2 percent. At this threshold value, the pattern of trees occupying the orchard is self-similar, from the overall pattern spanning the orchard down to the individual tree. Above the percolation threshold, the tree cluster is self-similar only on certain scales. On other scales, the pattern is merely some euclidean object.

Another intriguing and scientifically rewarding pursuit is to look at fractals upon fractals. Diffusion, as exemplified by the peregrinations of perfume molecules, is a fractal process. When diffusion occurs along a fractal surface, the process is something like the wanderings of an ant in a labyrinth *(see Figure 5.12)*. An ant constrained to roam along a straight line always returns to its starting point eventually. On a two-dimensional plane surface, the ant gets lost. But on fractal paths with dimensions between 1 and 2, what happens to the ant isn't completely known. Depending on the nature of its fractal labyrinth, the ant may keep running into dead ends forever, or it may return to its starting point very infrequently.

## STICKY STUFF

One of the wonders of nature is how simple water molecules can settle into symmetrical structures that have the limitless variety and intricacy of snowflakes. Very little is known about how these lacy shapes are created. Why does a snowflake branch? How does one branch tell another which way it's going so that the whole flake stays more or less symmetrical?

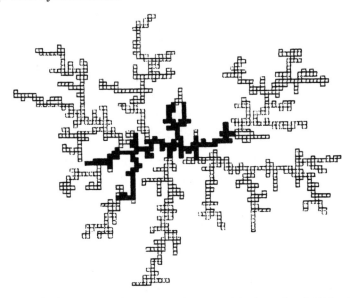

**FIGURE 5.12** Visited sites are shown in black in this 2,500-step random walk on a 1,000-site fractal substrate.

To try to account for how a snowflake or other growing body flowers into its final form, scientists have developed a variety of simple mathematical models that suggest ways in which growth can occur. By studying these models and comparing them with experimental observations, they can begin to guess what forces and conditions underlie various types of growth.

One of the simpler growth models, out of which a fractal form springs, starts as a tiny cluster, or "seed." Every time a particle wandering nearby happens to bump into the cluster, it sticks and stays put. Once it has arrived, the particle never jumps to another site. This type of process is called aggregation. If particles diffuse to the cluster by means of random walks—ordinary brownian motion in three dimensions—the process is known as diffusion-limited aggregation.

This growth process can be simulated in two dimensions, step-by-step on a computer. At the center of a grid sits a small, connected set of dots representing the initial aggregate. Each dot lies in a separate square of the grid. Far away from the central cluster of dots, a single particle starts its random walk. When the walker eventually arrives at an unoccupied space neighboring a square already filled, it stops and stays put, and the cluster grows by one unit. The process is continued for perhaps 50,000 or 100,000 wandering particles.

The resulting pattern is a typical fractal object, which looks like a bare tree seen from above, with branches shooting off in all directions (see Figure 5.13, left and Color Plate 7). This particular fractal object also contains tremendous empty regions, often in the form of long, narrow channels, or fiords, that penetrate far into its interior. What accounts for this distinctive structure is the fact that few late arrivals can get deep into a fiord. Random walkers that happen to reach a fiord's outlet and begin their journey down the channel are much more likely to bump into a wall than they are to make it all the way to the end.

In an effect called growth instability, any cluster that is slightly distorted initially will become even more distorted. Once the initial cluster begins to develop bumps and hollows, the bumps end up growing much faster than the hollows. Because particles are more likely to stick near peaks, the bumps grow even steeper, and the fiords become less and less likely to fill. Eventually, continual growth and splitting of the protruding ends give rise to a heavily branched, globular fractal.

What happens if a particle sometimes bounces off instead of sticking? Computer simulations show that this leads to a thickening of the branches and an increase in fractal dimension, but the figure remains a fractal.

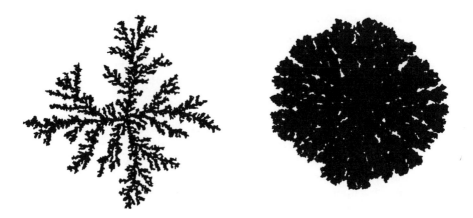

**FIGURE 5.13** *Left,* a 100,000-particle cluster grown on a square lattice using the diffusion-limited aggregation model; *right,* a 180,000-particle cluster formed in a two-dimensional simulation of ballistic aggregation.

Although diffusion-limited aggregation is easy to describe and simulate, the process is not yet well understood at a deeper level. No one is sure why the process gives rise to fractals rather than shapeless blobs that have no symmetry at all. Why are loops—or open lakes surrounded by matter—so rarely formed? How does the fractal dimension depend on the dimension of the space in which the process occurs? Ordinary mathematical tools seem inadequate for answering these and related questions.

The diffusion-limited aggregation model can be used, nevertheless, to explain a particular natural growth process, such as the deposition of metals during an electrochemical reaction *(see Figure 5.14).* Not only do many natural forms have fractal characteristics, but fractals can also serve as models for new materials that don't exist naturally. For example, diffusion-limited aggregation leads to materials with an especially large surface area. Experimentalists have created one such material by letting microscopic gold balls diffuse toward a small cluster, which grows into a three-dimensional aggregate that's so branched it's practically all surface.

What about snowflakes? Diffusion-limited aggregation theory has made a good start on reproducing the branched, hexagonal laciness of snowflakes. A sixfold pattern is easily built in when one lets the particles diffuse on a triangular lattice rather than a square grid. A controlled touch of randomness, or "noise," decides which of many equally probable sites should grow at each step. The resulting figures look like snowflakes, although they lack the amazing similarities shown among the branches of different arms in a real snowflake.

Another way to create a growth model is to begin a computer

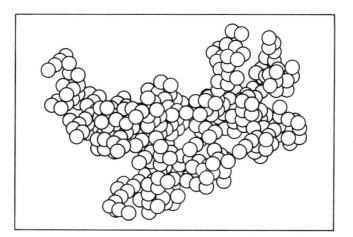

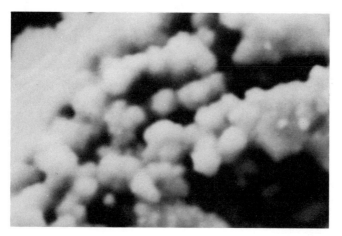

**FIGURE 5.14** *Top,* diffusion-limited aggregation, simulated in a three-dimensional space by a computer, gives rise to a fractal with a dimension 2.4; *bottom,* the pattern is similar to that of a copper cluster.

simulation with many particles distributed throughout space and to allow the particles to move about randomly until they meet and stick together. In this case the clusters can also move, which leads to a different type of pattern with a fractal dimension of about 1.4.

In ballistic growth, a given particle moves not on a random walk but along a randomly aimed straight line. If it strikes a predecessor, it sticks where it hits. The process generates patterns in the plane that after a few thousand trials look like somewhat porous, deckle-edged spots roughly circular in shape *(see Figure 5.13, right).* Simulations show that if growth goes on for long enough—over a quarter of a million or so trials—the laciness dwindles and the covered area grows with the square of the radius, while the edges become better defined. The fractal dimension approaches the euclidean value of 2.

But in most fractal growth models, mathematically precise results remain elusive because so many of their properties aren't known or understood. Even the concept of fractal dimension isn't enough to distinguish among the diverse objects that appear to be fractals. To oversome this deficiency, various other types of dimensions and constants have been introduced for special cases. Mathematicians and scientists have also been searching beyond the fractal dimension for other properties that may be universal.

A startling result of current research on fractal-modeled physical processes is that very different phenomena give rise to fractal dimensions that are very close in value. Such events range from the distribution of galaxies in the universe to the nature of turbulence in flowing fluids. It's too early yet to tell whether these phenomena are a collection of separate problems with a different explanation for each case or they depend on some underlying principle that would explain many of them simultaneously.

Meanwhile, the use of fractals as a descriptive tool is diffusing into more and more scientific fields, from cosmology to ecology. To some extent, fractals are in the eye of the beholder. Just as a river system can be described by naming it as a whole or by naming each of its numerous tributaries, researchers can choose various ways to describe the objects they see in nature. They can label the branches individually, or recognize that a branch looks roughly like the whole object and call it a fractal. Time will tell whether or not it's useful to see objects in terms of fractals. A meter stick is a very humble object until it gets into the hands of an Einstein. Einstein's work with meter sticks and clocks led to profound changes in our concepts of space and time. The fractal meter stick has yet to find its Einstein. Can this new meter stick be used to construct a deeper theory of why fractal objects occur, why certain fractal quantities are universal, and why fractals are ubiquitous in nature?

Ever since the earliest applications of fractal geometry, description has outpaced explanation. Theories that depend on fractal properties seem to work, but no one really knows why. Further progress in the field depends on establishing a more substantial theoretical base in which geometrical form can be deduced from the mechanisms that produce it. Without a strong theoretical underpinning, much of the work on fractals seems superficial and perhaps pointless. It's easy to perform computer simulations on all kinds of models and to compare the results with each other and with experimental results. But without organizing principles, the field drifts into a zoology of interesting specimens and facile classifications.

# 6

## THE
## DRAGONS
## OF
## CHAOS

With barely any warning, the human brain can fail. When that happens, a victim may spend seconds staring blankly into space or, in extreme cases, lose consciousness and fall stiffly to the ground while the entire body shakes. Such seizures, or epilepsies, are symptoms of uncontrolled overactivity among the brain's nerve cells. In some way, an electrical disturbance originating in a few neurons temporarily takes over the brain, and no other messages get through.

Mathematical techniques now being developed may provide some insight into the causes of seizures. The same mathematical methods can also be applied to complex computer systems and networks with numerous interconnections and a wide range of independent but coordinated functions. These mathematical models predict that just as a small group of neurons may suddenly generate a disruptive electrical signal, so too can electronic components within a computer.

## STEPS IN TIME

The still-tenuous thread that ties together computer networks and physiological systems is an emerging field of mathematics called *topological dynamics*, which describes the way in which systems change over time. This mathematical approach suggests that systems governed by physical laws can undergo transitions to a highly irregular form of behavior now termed *chaos*. Although chaotic behavior appears random, it is governed by strict mathematical conditions.

Like many terms appropriated by scientists to encapsulate a new theoretical concept, the word "chaos" takes on a specific meaning in a scientific context. The technical definition of chaos, somewhat tenuously related to its everyday meaning, carries with it an image of order in the midst of disorder. That definition contrasts with normal usage, wherein we mean by chaos either a state of utter confusion or a state in which chance is supreme.

Electronic and biological systems are not the only places where chaos is likely to strike. It shows up in many physical systems, especially those that involve turbulence — from the way in which water in a mountain stream swirls and splashes down a rocky channel to the way smoke rising in still air starts as a well-defined plume, then gradually breaks up into a disorderly tangle of smoky threads.

Some motions, such as the trajectory of a baseball, are inherently predictable. A fielder intuitively knows where the ball is likely to fall and automatically uses that information to catch it. On the other hand, the path of a flying balloon expelling air is erratic and unpredictable. The balloon lurches and turns at times and places impossible to predict. Nevertheless, both balloon and baseball obey Newton's laws of motion. Somehow, predictability and unpredictability together reside in the same set of equations. This contradicts the old notion that given the initial position and velocity of each particle in a system and applying newtonian mechanics, the system's history can be predicted forever. The idea works reasonably well for planets in the solar system but doesn't seem to apply to the way ocean waves crash into a beach or to the "noise" encountered in an electronic circuit.

Understanding chaotic behavior usually means tracking the way a system behaves over time. It also means closely examining the solutions of differential equations that describe and model the phenomena that scientists encounter in nature. In mathematical models of neural systems, the variables could represent, for example, the electrical potential across cell membranes, membrane currents, and the concentrations of the chemical substances present. Some parameters, such as temperature or a particular drug concentration, assume a constant value for a given case but may vary from case to case.

A differential equation links the rate of change over time of one of the variables with the variable's current size and the current size of other variables. Descriptions of complicated systems may require a large set of such equations. In a sense, a set of differential equations is like a machine that takes in values for all the variables and then generates the new values at some later time. Often, the relationship expressed in the equations is nonlinear; that is, input and output are not proportional.

Mathematicians have learned that under the right conditions, even simple sets of nonlinear differential equations can yield numbers that appear to follow no pattern. Although the equations express direct cause and effect relationships, the numerical results predict that modeled systems can show irregular motion or randomlike, chaotic behavior. In fact, this class of solutions displays a sensitive dependence on initial conditions, so that a slightly different starting point produces a radically different result. In principle, the future is completely determined by the past, but in practice, small uncertainties are amplified, so that even though the behavior is predictable in the short term, it is unpredictable in the long term.

Such irregularity can have startling consequences. If weather systems can be described by mathematical equations that shift into chaotic behavior, then a change as slight as a butterfly flapping its wings near a weather station would make long-term weather predictions impossible. In general, the existence of chaotic regimes implies new fundamental limits on predictability, if not a theoretical excuse for the notorious unreliability of local weather forecasts today. At the same time, the determinism inherent in chaos suggests that many random phenomena are more predictable than had been thought.

The mathematical side of chaos is seen most easily in phase space, where, as we saw in Chapter 4, each dimension represents one of the variables in the differential equations used to model a particular system. Researchers are interested in what happens to phase-space trajectories for different equations and under various circumstances. Sometimes the motion, as seen in phase space, settles down and is sucked into one point and stays there. That fixed point is an example of an *attractor*. A mass oscillating on the end of a spring, for example, gradually loses its energy and stops. Its phase-space portrait, with position plotted against velocity, looks like a spiral that gets tighter and tighter, ending up at a fixed point. If there were no friction, the spring's motion would continue forever, and its phase-space portrait would be a single loop along which the motion endlessly circulates.

A pendulum clock, in which energy lost to friction is replaced by a mainspring or a system of falling weights, cycles periodically through a sequence of states. The same principle applies to metronomes and the human heart: they repeat the same motion over and over again. In phase space, such a motion corresponds to a cycle or a periodic orbit. Such attractors are called *limit cycles*. No matter how the motion is initiated and despite initial fluctuations, the cycle approaches the same long-term limit.

A particular system may also have several attractors. Different initial conditions drive the system to a different attractor. The set of points going to the same attractor is called a basin of attraction. Attractors themselves can have a variety of geometric shapes, from something as simple as a torus to complicated, higher-dimensional forms. Nevertheless, motion on all the attractors discussed to this point is completely predictable, so it is nonchaotic.

Under certain conditions, however, nonlinear differential equations generate trajectories in phase space that form peculiar shapes, having none of the regularity associated with the previous examples of attractors. Such objects are called *chaotic*, or *strange*, *attractors*. Trajectories skip about on the surfaces completely unpredictably. A slight change in the initial conditions greatly changes the particular orbit generated.

In 1963, meteorologist Edward Lorenz of the Massachusetts Institute of Technology discovered a strange attractor in a simplified set of differential equations describing air flows in the atmosphere. That attractor is now known as Lorenz's butterfly, or mask, because its curious shape resembles the flapping wings of a butterfly or the kind of mask worn by the Lone Ranger *(see Figure 6.1)*. As with all other strange attractors, the behavior of a trajectory on its surface is totally erratic and unpredictable, but the trajectory never leaves the attractor's surface. In weather terms, Lorenz's attractor places limits on what the weather can do. For example, it's not possible to have a blizzard in the Sahara Desert or a 200-degree heat wave in New York City. But within certain limits and given enough time, the weather

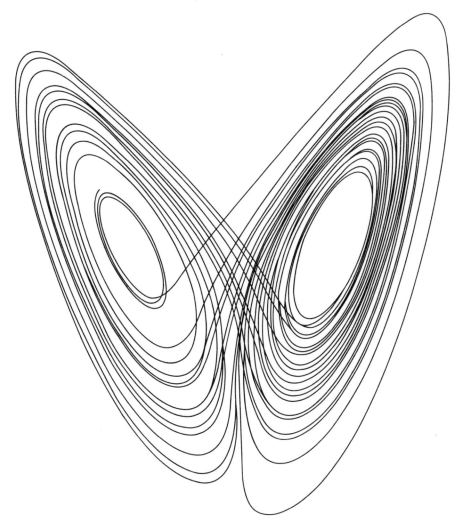

**F I G U R E  6.1**     A set of nonlinear differential equations for modeling air flow in the atmosphere generates a butterfly-shaped figure now called the Lorenz attractor.

can do practically anything. No matter how many data on temperature, barometric pressure, and wind direction are collected, the weather's real behavior over time veers dramatically from predictability.

Since Lorenz's discovery, mathematicians and other researchers have discovered more strange attractors, which crop up in various nonlinear differential equations. Chaotic attractors have also been observed in several experiments related to fluid flows and chemical reactions. Among them are the convection pattern of fluid heated in a small box, oscillating concentration levels in a stirred chemical reaction, the beating of chicken-heart cells, and a large number of electrical and mechanical oscillators. In addition, computer models of events ranging from disease epidemics to electrical activity of a nerve cell to stellar oscillations have expressed this simple type of randomness. There are even experiments now underway that are searching for chaos in areas as disparate as brain waves and price fluctuations.

Further study of both simulated and observed chaotic attractors shows that they are fractals, usually of a dimension greater than 2. As we saw in Chapter 5, magnifying a fractal shows more detail, no matter how much magnification is done. Furthermore, chaotic attractors act like a pump to bring microscopic fluctuations to a macroscopic expression: small-scale uncertainties are made larger and larger. Because the starting point can never be known with sufficient precision, no exact solution — and no mathematical or theoretical shortcut to predicting the future — is possible.

This curious consequence implies that prediction isn't necessarily a good test of a theory. The classical approach to verifying a theory is to make predictions and test them against experimental data. If the phenomena are chaotic, however, long-term predictions are intrinsically impossible. This has to be taken into account in judging the merits of a theory. The process of verifying a theory thus becomes a delicate operation that relies on statistical and geometric properties rather than on detailed prediction.

One important insight that these dynamical studies provide is that systems may behave "normally" for a wide range of initial conditions and then suddenly shift into a chaotic mode when one parameter, such as the concentration of an administered drug in the case of a neural system, moves through a critical value. Thus, a tiny change in parameter can result in dramatically altered behavior. This abrupt change in behavior, obtained reproducibly in response to a small change in the value of a system parameter, is, in mathematical terms, a *bifurcation*, or branching.

In the case of epilepsy, researchers are exploring whether transitions from normal behavior to convulsions are the result of bifurcations. Such knowledge would greatly facilitate treatment. A reliable mathematical model would allow physicians to identify which drugs reset parameters so that a convulsion ceases.

Mathematical methods are now available that permit the analysis of dramatically irregular behavior. Highly disordered behavior no longer need be viewed as uncontrollable or inevitable. It can, at least, be analyzed by matching the appropriate mathematical model with the right physical situation.

# ——— THE STRANGER SIDE OF SQUARES ———

The penetrating squeal that escapes from a loudspeaker when a microphone picks up the speaker's sound is a striking example of feedback. The output from the microphone and amplifier is continually cycled back into the system as new input. Rapidly, the whole process gets out of hand.

Mathematics has its own version of feedback, and its results are often equally startling. In this case, the "amplifier" is an algebraic expression, such as $4x(1 - x)$. Computing the value of that expression for some initial value of $x$, then substituting this answer back into the original expression starts the loop. Repeating this simple, iterative process over and over again leads to surprisingly complex, even unpredictable, mathematical behavior. Each step, though completely determined by the equation, may take a traveler on an unexpectedly erratic course. This mathematical phenomenon expresses the same kind of disorder that arises with nonlinear differential equations. But in this instance, the equations are exceedingly straightforward.

A simple calculator experiment illustrates what may happen. The sequence of numbers 0.2, 0.64, 0.922, 0.289, 0.822, 0.586, 0.970, 0.116, 0.406, . . . looks haphazard, yet it's the result of starting with the number 0.2 and substituting that value for $x$ in the equation $4x(1 - x)$. This gives the second number in the sequence, which is 0.64. Substituting $x = 0.64$ into the same expression gives the next answer, 0.922, and so on.

Further exploration in this astonishingly chaotic regime turns up another surprise. A slightly different starting value leads to a sequence that bears little resemblance to one initiated by a near neighbor. For the seed value 0.21, for instance, the sequence is 0.21, 0.664, 0.892, 0.364, 0.926, 0.274, 0.796, 0.650, 0.910, . . . , a far cry from the sequence starting at 0.20.

A calculator experiment does not, however, provide conclusive evidence that something unusual is going on. The results may be affected by rounding off answers to three decimal places; the chosen examples may be special cases; or the iteration may not have been carried out far enough. What's needed is a more thorough, systematic look at the equations and the process. With some equations, nothing much happens. For a linear expression such as $3x$, each iteration simply triples the initial value of $x$, no matter what the initial value is. The numbers in the sequence get larger and larger in predictable steps. Iterating something like $x^3$ also results in values that keep going up and up or down and down. In fact, after $n$ steps, the $n$th value in the sequence is easily determined: It equals $x^{3^n}$. Clearly, to see cyclic or erratic behavior, the mathematical expression must have some sort of switchback or twist.

The simplest example of such a feedback system is represented by the nonlinear equation $x_{n+1} = kx_n(1 - x_n)$. This equation determines the future value of the variable $x$, at time step $n + 1$, from the past value of $x$ at time step $n$. Known as the *logistic equation*, it is often used in ecology as a model for predicting population growth. The variable $x$ can be thought of as the population of, say, gypsy moths at a particular time. The equation expresses how the population in a given year depends on the population in the previous year. The parameter $k$ can be adjusted to reflect a variety of biological constraints. An enormous amount of work has gone into understanding the behavior of sequences defined by this equation, over a wide range of initial conditions and parameter values.

When the logistic equation is written in form $x_{n+1} = kx_n - kx_n^2$ its true mathematical identity becomes clearer. This is a quadratic equation with a linear first term and a nonlinear (quadratic, in this case) second term. The relative importance of the two terms depends on the chosen starting point. For example, when $x_0 = 0.01$, then $x_0^2$ is a miniscule 0.0001. Thus, if $x_0$ is sufficiently small, then the nonlinear term is even smaller and barely affects the computation. In this situation, the population rises steadily when $k$ is greater than 1 and falls steadily when $k$ is less than 1. In other words, the linear term can be interpreted as a birth or death rate, leading to exponential growth or decay. On the other hand, when $x_0$ is large, the nonlinear, or feedback, term becomes important. Because this term is negative, it represents a death rate that dominates when the population gets too large. Biologically, that nonlinear death rate could be the result of food supply shortages, widespread disease, or other factors that become important in an overcrowded environment.

For a given value of the parameter $k$, once a starting point $x_0$ is specified, the evolution of the system is fully determined. One step inexorably leads to the next. The whole process can be pictured on a graph—a sort of guide map—on which the path from one value of $x$ to the next can be traced. Indeed, such a graph is called a *return map*.

For the logistic equation, plotting $x_{n+1}$ against $x_n$ results in a parabola that opens downward like an upside-down bowl. The inverted parabola sits on the $x_n$ axis, wedged between 0 and 1. Its rounded peak reaches a maximum value of $x_{n+1} = k/4$ when $x_n = 0.5$. With this graph, a logistic explorer can track the equation's numerical evolution, so long as the starting point is between 0 and 1 on the $x_n$ axis.

The simplest course is to locate a starting point $x_0$ on the $x_n$ axis, then shoot vertically upward to intersect the parabola. A horizontal glance from the parabola toward the $x_{n+1}$ axis provides the value $x_1$. Returning with $x_1$ as a new starting point on the horizontal ($x_n$) axis allows the whole process to be repeated to obtain $x_2$.

This graphical procedure seems to be almost as much trouble as computing each of the numbers in a sequence, but a handy shortcut simplifies the process: all it takes is the addition of a slanted line 45 degrees up from the horizontal axis (representing the line $x_{n+1} = x_n$). Now the best course to steer is from $x_0$ vertically to the parabola to reach $x_1$, then horizontally to the 45-degree line, and vertically back to the parabola for $x_2$, and so on. Each step, or iteration, requires a horizontal move to the 45-degree line, then a vertical jump or drop to the parabola *(see Figure 6.2)*.

These paths, or *orbits*, give the first indication of which routes lead to the erratic behavior of chaos. Whereas some orbits converge on one particular value, others jump back and forth among a few possible values, and many roam, never settling anywhere. Exactly what happens depends strongly on the value of the parameter $k$, which measures the degree of nonlinearity. As $k$ increases, going from 0 to 4, the logistic equation shows an amazing transformation from order to chaos.

The transformation happens in clearly defined stages. When $k$ is less than 1, every starting point produces a path that eventually ends up at 0. When $k$ is between 1 and 3, just about every route, no matter where it starts, is eventually attracted to a specific value called a *fixed point*, which occurs where the parabola intersects the 45-degree line at $x = (k - 1)/k$. This corresponds to a steady-state, or equilibrium, population, which occurs when the numbers of births and deaths balance.

When $k$ is larger than 3, the fixed point becomes unstable, and wild and woolly things begin to occur. At $k = 3.2$, two fixed points appear, and the value of $x$ oscillates between two values of $x_n$, roughly equal to 0.5 and to 0.8. In this case, the orbit for any starting point is

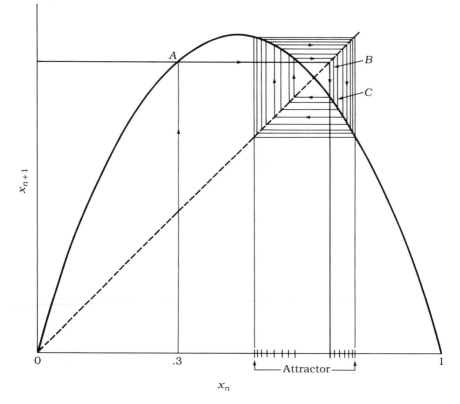

**F I G U R E  6.2**    The logistic map defines an inverted parabola. Once the initial value is specified, values of $x$ at succeeding time steps can be found by tracing a path vertically up to the parabola at point $A$, horizontally to point $B$ on the 45-degree line, vertically down to point $C$ on the parabola, and so on. Eventually, the path settles down into a cycle, going back and forth between two values, which comprise the attractor.

trapped in a period-2 cycle; that is, it skips relentlessly from one to the other value. Boosting the value of $k$ adds two more fixed points, and the orbit visits all four values in turn, returning to the original point in four time steps. That behavior puts these orbits in a period-4 cycle. As $k$ grows, the period and the number of fixed points double and redouble, until at $k$ approximately equal to 3.57, both the period and the number of fixed points are essentially infinite. The trajectories for practically all starting points appear to skip about restlessly. For $k$ greater than 3.57, no patterns appear in the sequence of successive values for $x$. The abrupt population changes evident for these values of $k$ are indistinguishable from a random process, although there are no random forces involved, and, according to the logistic equation, the future is strictly determined by the starting point $x_0$.

The cascading, period-doubling route to chaos is itself worth exploring. The range over which a particular cycle is stable decreases quickly as the cycle's period and the number of fixed points increase. Longer and longer periods are crowded into less and less space. In fact, physicist Mitchell J. Feigenbaum, having observed this period-doubling sequence in numerical experiments on a computer, was able to prove that the successive stable intervals become smaller in a very particular way: the range of values for $k$ over which the period-2 cycle is stable is about 4.6692016 . . . times wider than the range for the period-4 cycle; the range for the period-4 cycle is the same amount wider than the period-8 cycle, and so on, leading to the rapid accumulation of cycles with longer and longer periods until the limit at infinity is reached.

To get a better idea of what is happening as periods increase and chaos begins to take over, we need another graph. Plotting successive values of $x$ on one axis against the parameter $k$ on the other, as $k$ changes from 3.5 to 4, provides a broad canvas against which the logistic equation's behavior can be viewed. Each return map for a particular value of $k$, iterated hundreds of times, is compressed into a narrow vertical strip. Piecing together these strips in a wide band produces the computer-generated mural known as a bifurcation diagram (see Figure 6.3).

The result is a graphic display of the underlying structure of chaos. Forking patterns show an orderly progression from regular to chaotic behavior. Wherever the long-term behavior of orbits for a given value of $k$ converge to a cycle of, say, period 4, the diagram shows 4 discrete values for $x$. For $k = 3.5$, the system eventually settles down to a periodic oscillation among these four values. Where the evolution is chaotic, the values of $x$ seem to cover the whole interval, leaving a dark strip. But even the chaotic domain is criss-crossed by dark streaks, showing that some $x$ values are visited more often than others. Intersecting streaks signal mathematical crises, where nearby orbits are forced to wander in a tight space. A particularly dramatic collision is visible at $k = 3.68$.

Windows of periodic behavior also sit embedded in the chaotic regime. A period-3 cycle, for example, shows up at $k$ roughly equal to 3.83. Here the population increases in two successive years and decreases in the third. As $k$ edges upward, that cycle becomes unstable, the period doubles, and the number of fixed points doubles to 6, then to 12, then 24. It turns out that there are windows of stability for cycles of all whole-number periods, each of which exhibits a cascade

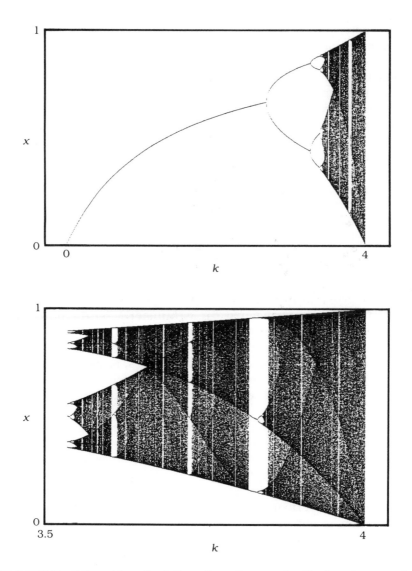

**FIGURE 6.3** *Top,* the bifurcation diagram for the logistic map; *bottom,* a detailed look at the region between $k = 3.5$ and $k = 4.0$, when chaos makes its appearance.

of period-doubling bifurcations back to chaos. Yet despite the infinite number of such intervals of stability throughout the range of $k$, there's still plenty of room for truly chaotic motion. In other words, if the exact evolution of $x_n$ looks chaotic, then it probably is. It's unlikely to be evidence of very long but periodic cycles. By the time

$k = 4$ — and $x$ can end up anywhere between 0 and 1 — it's relatively easy to show that the orbits meet the definitions of both a chaotic and a random process.

At first glance, the complexity of the dynamics shown by even a simple expression such as the logistic equation is somewhat dismaying. It's like stepping into a modest cottage and finding a veritable palace inside, full of secret passages and mysterious rooms. What happens with more complicated equations? Luckily, features like the Feigenbaum constant and period-doubling bifurcation sequences appear in a variety of systems *(see Color Plate 8)*. Whole classes of equations, when iterated, exhibit exactly the same behavior as quadratic polynomials. These features are universal, in the sense that they depend solely on the presence of feedback and are virtually insensitive to other system details.

Feigenbaum's period-doubling route to chaos has also been observed in physical systems as diverse as turbulent fluids, oscillating chemical reactions, nonlinear electrical circuits, and ring lasers. The logistic equation itself seems to mirror well the wild and unpredictable fluctuations that may occur in gypsy moth populations, in stock and commodity prices, and in the dynamics of some mechanical oscillators.

## ——— THE COMPUTER AS MICROSCOPE ———

There are few more striking demonstrations of the complexity hidden in simple laws or rules than the multitude of structures contained within the Mandelbrot set, which was briefly introduced in Chapter 5. This figure is truly one of the most complicated objects in mathematics. From within its bounds and around its borders come pictures of order and chaos, conflict and coexistence, and transition. Though often enigmatically beautiful, these mathematical products of computer graphics also represent an attempt to understand how subtle changes in equations affect their behavior.

The polynomial $x^2 - 1$, under iteration, does nothing particularly exciting. From any starting point defined by an ordinary (or "real") number, successive values show a humdrum predictability. The fireworks begin when complex numbers are used in place of real numbers. With complex numbers, what was only a crack in a wall becomes a full-fledged picture window, revealing a truly wondrous, chaotic landscape.

Real numbers can be thought of as labels for all the points on a line; for example, the real number 1.5 lies halfway between points 1 and 2, and 1.67 lies somewhere between 1.5 and 2. Every real number has its place along the number line, and every point on the number line has its own real-number label. Complex numbers, on the other hand, define a point not on a line but somewhere in a vast two-dimensional sheet of numbers called the *complex plane.*

A complex number, *z*, has two parts, which for historical reasons are called real and imaginary. These terms no longer have a literal meaning; the two parts could just as easily be named Tweedledum and Tweedledee, or Albert and Brenda. A complex number may be written in the form $a + bi$, where *a* is the real part and *bi* is the imaginary part. The italic *i* next to the *b* shows which part of the complex number is imaginary. In this notation, $4 + 3i$, $0.5 + 1.76i$, and $16 - 8i$ are all examples of complex numbers.

Every complex number can be represented by a point in the plane. Locating a particular complex number requires defining two axes: a horizontal real axis and a vertical imaginary axis. Assuming the axes intersect at the point $0 + 0i$, then the number $4 + 3i$, for instance, would lie 4 units to the right and 3 units vertically upward from the origin where the axes intersect. This point would be 5 units away from the origin. The complex plane is filled with uncountably many such pairs of numbers. Their real and imaginary parts can be either positive or negative and either whole numbers or decimal expansions.

Adding two complex numbers is easy. To find the sum of $4 + 3i$ and $2 - 7i$, the real and imaginary parts are added separately. In this case, the sum is $6 - 4i$. Under addition, the two types of terms stay segregated.

Multiplying two complex numbers is only slightly more difficult. If the symbol *i* is treated like an *x* in algebra, then the product of $4 + 3i$ and $2 - 6i$ is $8 - 24i + 6i - 18i^2$. It happens that $i^2 = -1$. The product now becomes $8 - 24i + 6i + 18$. Simplifying by collecting the real and imaginary parts reveals the final answer: $26 - 18i$.

With rules for addition and multiplication and a vast plane to explore, a new journey in search of chaos and complexity can begin. The same questions that come up for expressions involving real numbers also come up for expressions involving complex numbers. Will the iterative process, starting at an arbitrarily chosen value, generate a sequence of numbers that get closer and closer to a particular value and eventually come to rest there? Will the sequence arrive at a cycle of values, repeated over and over again? Or is the sequence erratic and unpredictable?

For the expression $z^2 - 1$, some choices of starting values generate sequences that grow without bound, flying off to infinity. Other starting values lead to sequences that stay within a well-defined boundary on the plane. This boundary has the fractal structure reminiscent of a deformed island coastline spotted with wartlike protuberances and pinched into deep bays *(see Figure 6.4)*. Magnifying a small piece of the border reveals details that mimic the overall raggedness of the entire border.

This kind of boundary is known as a *Julia set*, named for French mathematician Gaston Julia, who, along with Pierre Fatou, first studied these forms in the early part of this century. They established the idea that the entire boundary could be regenerated from an abitrarily small piece of the boundary. Their fascinating work, however, remained largely unknown, even to most mathematicians, because Julia and Fatou lacked the tools needed to communicate their subtle ideas. Now high-resolution computer graphics brings their visions to life.

Different quadratic equations lead to different pictures of the behavior of sequences. One particular set of equations can be written as $z^2 + c$, where $c$ is a given complex number. Any computer can perform the necessary iteration: square the starting number, then add the constant to generate a new number to be squared, and so on, over and over again. Because complex numbers represent the coordinates of points in two dimensions, each iteration can be viewed as a hop from one point in the plane to another. As in the case of real numbers, the set of hops can be thought of as a trajectory or an orbit.

When $c = 0$, three types of behavior are possible. All starting points that lie within a distance of 1 of the origin are drawn toward

**FIGURE 6.4** Iterating the expression $z^2 - 1$ generates a Julia set, in which the black area represents the set of complex-number starting points that remain bounded.

the origin. It's as if a magnet, of strictly limited power, were sitting at the point $0 + 0i$, attracting any hopping particles that happen to start within its circle of influence. In fact, zero is said to be an attractor for the process. Successive squarings yield orbits that tend to zero.

All points more than a distance of 1 from the origin end up sailing farther and farther away. Somewhere at infinity, another magnetic attractor is doing its job. As the iteration continues, the numbers steadily become larger. Points on the boundary, a circle of radius 1, stay where they are, caught in the competition between the two attractors. Overall, two zones of influence divide up the plane, and the boundary between them is simply a circle. This boundary is an especially simple Julia set.

Varying $c$ leads to an infinite number of different pictures. For $c = -0.12375 + 0.56508i$, the central attractor is no longer zero, and its outer boundary—the limit of its circle of influence—is no longer smooth but crumpled. When $c = 0 + 1i$ (or just plain $i$), an orbit starting at 0 is drawn into a never-ending oscillation between the two complex numbers $-1 + 1i$ and $0 - 1i$. For other starting points, the sequence may also be attracted to two fixed points, or it may explode to larger and larger values. Picking $c = -0.12 + 0.74i$ yields not a single island of attraction but an infinite number of barely connected, deformed circles.

Clearly, by varying $c$, an incredible variety of Julia sets can be generated. Some look like fat clouds; others are like twisted, thorny bushes; many look like sparks floating in the air after a flare has gone off. With a little imagination, a person flipping through a book of these Julia sets may glimpse the shape of a rabbit, the jagged form of a dragon, and the curls of a sea horse. Sometimes, the Julia set is merely a dust of disconnected points (*see Figure 6.5*).

Indeed, the book of Julia sets for iterations of the equation $z^2 + c$ has an infinite number of pages, one for each possible value of $c$—which covers every point in the complex plane! When mathematicians began to study all these forms, they soon realized that they needed a kind of table of contents that would give them an overview of this huge book filled with page after page of incredibly complex figures. That organizing principle was discovered in 1980 by Benoit Mandelbrot, and it now carries his name: the *Mandelbrot set*.

In every case, depending on their starting points, some orbits stay bounded no matter how long their hops are tracked, whereas others streak off to infinity. This difference in behavior makes the boundary between these two regions of crucial importance in defining the properties of a given equation.

It turns out there are only two major types of Julia sets, no matter what $c$ is. Either the area within a boundary is all one piece (that is,

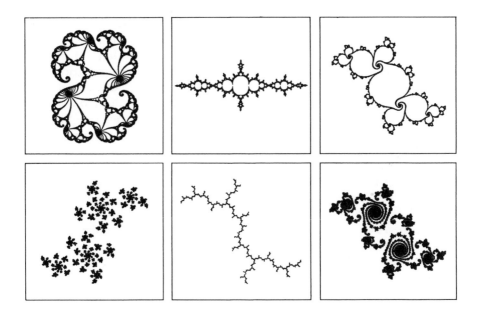

**FIGURE 6.5**    Julia sets, when iterated for various values of $c$ in the expression $z^2 + c$ come in many forms, some connected and some fragmented.

the area is a connected structure), or it's broken into an infinite number of separate pieces to form a cloud of points, something called a Cantor set, introduced in Chapter 5. No matter where anyone looks in the book of Julia sets, each page falls into either the first or the second category. The Mandelbrot set consists of all values of $c$ that have connected Julia sets.

That definition makes it possible to draw a portrait of the Mandelbrot set. If the Julia set corresponding to a certain complex number $c$ is a cloud of dust, that point in the complex plane is colored white. If the Julia set is connected, the point is colored black.

If the only way to draw the Mandelbrot figure meant looking at all the different Julia sets and deciding which ones belong and which ones don't, then it would take an eternity to compile the table of contents. But mathematicians John Hubbard and Adrien Douady found a quick way to generate the set. They proved that it's enough to know for a given value of $c$ whether the starting point 0 stays bounded. If it does, then the point $c$ belongs to the Mandelbrot set. Hubbard describes his procedure, evaluated for all values of $c$, as probably the most economical definition yet known of a complex object. A computer program written to generate the Mandelbrot set need be no more than four or five lines long.

Finding the set comes down to computing the sequence: $c$, $c^2 + c$, $(c^2 + c)^2 + c$, . . . and seeing where it goes. For instance, $c = -1 + 0i$ (or $-1$) is in the Mandelbrot set because its sequence hops back and forth between 0 and $-1$. The point $1 + 0i$, on the other hand, goes from 0 to 1 to 2 to 5, continuing ever upward, fleeing toward infinity.

The rotund, wart-covered, black figure that emerges from evaluating Hubbard's sequence is, in effect, the required one-page table of contents (see Figure 6.6). This awesome image says something about all quadratic equations at once, providing precise information about every page in the book of iterated squares.

The figure itself, looking like a pimply, squat snowman lying on its side, takes up only a small piece of the entire complex plane. Its pointed cap reaches to $-2$ along the real axis, while its indented bottom rests at $\frac{1}{4}$. Its arms extend out along the imaginary axis to $i$ and $-i$.

This strange, solitary figure is a source of unending wonder. Mathematicians such as Hubbard, Heinz-Otto Peitgen, and John Milnor have been unable to resist its allure and have spent a great deal of

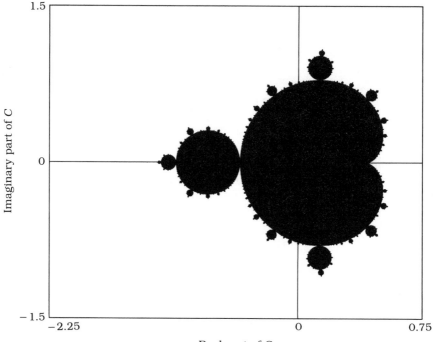

**FIGURE 6.6** The Mandelbrot set (shown in black) extends from the cusp of the cardioid at $c = 0.25$ to the tip of its tail at $c = 2$.

computer time exploring the figure's intricacies, developing and proving mathematical conjectures in the course of their computer-aided explorations. They have also constructed a considerable library of magnificent graphic images, portraits of the Mandelbrot set in all its moods.

The Mandelbrot set has detailed structure on all scales. Zooming in for a closer look shows fine structures that seem to get more and more fantastic and complicated as the magnification increases. Close-ups of its borders unveil a riot of tendrils and curlicues, yet everything is connected. A bewildering array of delicate filaments holds all the parts together *(see Figure 6.7)*.

Delving deeper and deeper also turns up miniature snowmen.

**FIGURE 6.7** A blowup of a piece of the Mandelbrot set's boundary reveals a riot of filaments and small copies of the Mandelbrot set.

Little Mandelbrot figures lurk wherever the computer peeks. For example, the Mandelbrot set wears a coat of fine, crumpled, branched antennae. A close inspection reveals that these antennae also carry many little copies of the larger Mandelbrot set, sitting like fuzzy, round marshmallows strung out on a skewer. Further magnification turns up even tinier copies of the Mandelbrot set. In fact, the Mandelbrot set includes an infinite number of copies of itself, all tied to one another by tiny lifelines. The details are always different, however, depending on the magnification and the figure's precise location.

Remarkably, each point in the Mandelbrot set carries its own address. It's possible to figure out where a blown-up region fits just by counting the number of strands that come together at various nearby points. Moreover, the Mandelbrot set does much more than an ordinary table of contents. It not only catalogs where everything is, but contains a miniature of each of the book's pages in an incredibly compressed form. Looking at a close-up of a tiny piece of the Mandelbrot set provides a glimpse of the corresponding Julia sets associated with those values of $c$ (see Figure 6.8).

If $c$ happens to lie well within the main body of the Mandelbrot set, the Julia set is a crinkled circle, surrounding a single attractor. If $c$ is inside one of the snowman's buds, the Julia set consists of infinitely many fractally deformed circles. Picking a value of $c$ from one of the set's branched antennae reveals a similarly furry Julia set. Every neighborhood has its own distinctive character.

Following a path that starts deep within the Mandelbrot set then crosses its boundary reveals a dramatic, fundamental change in the nature of the associated Julia sets. At the very edge of the Mandelbrot set, the Julia sets explode, splintering and falling to dust. As the path winds into distant parts of the complex plane, far away from the Mandelbrot figure, the scenery gets more sparse. The clouds of points in the Julia sets get thinner and thinner.

The greatest diversity flourishes in the border zone, where many cultures and customs are apt to mingle. This tangled region harbors a fantastic, baroque coterie of dragons, sea horses, and other strange creatures, contrasting sharply with the simplicity of the single mathematical expression responsible for the myriad forms that live in the zone. Attractors compete for influence on the plane. Every conflict breaks up into innumerable smaller battles.

The true complexity of the Mandelbrot set's border region is hard to grasp without the use of computer graphics. A computer can create a colored halo around the figure by calculating the escape time for each point on the screen. Points within the Mandelbrot set don't escape at all, whereas points outside escape at varying rates. Each point is colored according to a chart stored in the computer. The

**FIGURE 6.8** The Mandelbrot set determines the structure of corresponding Julia sets. Points that fall within the Mandelbrot set represent connected Julia sets, whereas those outside the Mandelbrot set are matched with fragmented Julia sets.

gradation of colors indicates how quickly escape occurs, quantifying the escape toward an attractor located an infinite distance away *(see Color Plate 9)*.

The incredible complexity of the Mandelbrot set provides a wonderful world for exploration, with beauty spots to visit, innumerable strange sights to view, and unexpected turns to any journey. In place of the Taj Mahal or the Eiffel Tower, the Mandelbrot tourist visits the unnamed local wonders marked by coordinates in the complex plane. Explore the area where the real part lies between 0.26 and 0.27 and the imaginary part between 0 and 0.01, the tour guide suggests. Or why not try $-0.76$ to $-0.74$ and $0.01i$ to $0.03\ i$; $-1.26$ to $-1.24$ and $0.01i$ to $0.03\ i$.

At the same time, the discovery of the Mandelbrot set is somewhat disquieting. If a mathematical expression as innocuous as a quadratic equation, iterated in the complex plane, exhibits such complexity, what hope is there of understanding more complicated equations? Fortunately, the Mandelbrot set, in various forms, shows up again and again in more complicated systems *(see Color Plate 10)*. Little by little, mathematicians are building a new picture of iterated mathematical expressions.

## ZEROING IN ON CHAOS

Engineers and scientists spend a lot of time solving equations. Such exercises provide the information needed to understand how molecules stick together, to determine the strength a steel framework must have to support a given structure, or to ascertain the size of distant galaxies. Equations are the engines that drive much of engineering and scientific work. Over the centuries, mathematicians have developed a variety of methods for solving equations. Now, using the capabilities of modern computers, they are starting to explore in detail how the age-old methods work: when they can be relied upon, when they fail, and when they behave strangely.

One calculus-based, equation-solving technique that has undergone intense scrutiny is *Newton's method*, which has been in use for centuries and is often taught in beginning calculus courses. Several mathematicians have recently probed this method's idiosyncracies and have come up with startling pictures of its behavior. Their results reveal a chaotic side to Newton's method that has become apparent only in the last decade. Earlier mathematicians had suspected that problems with the method could arise under certain circumstances, but lacking computer graphics, they could not express their concerns.

At the center of Newton's method (and much of calculus) is the concept of a function. A *function* is a rule that assigns a certain output to any given input. The squaring function is a familiar example: input 2, output 4; input 9, output 81. The reciprocal function turns 2 into $\frac{1}{2}$, and 5 into $\frac{1}{5}$. Other well-known functions include the sine and exponential functions.

More formally, a function's output can be written in mathematical shorthand as $f(x)$. Thus, the squaring function can be expressed as $f(x) = x^2$. When the input, $x$, is 2, the output, $f(x)$, is 4. When $f(x)$ is given as $x^2 + x - 6$, then $f(x)$ has a value $-4$ if $x = 1$. The process can also be reversed. Sometimes, the value of the function is known and the input value or values are unknown. That requires solving the equation; that is, finding values of $x$ that will make the equation true.

Anyone who studied high-school mathematics is probably familiar with the problem of finding values of $x$ for which, say, $x^2 + x - 6 = 0$. In this case, $x$ can be either 2 or $-3$. The two solutions, 2 and $-3$, are known as the roots of the equation.

One way to picture what is happening is to plot a graph that shows what the function $x^2 + x - 6$ looks like. When the function's value, $f(x)$, is computed for various values of $x$, the resulting pairs of numbers represent coordinates that can be plotted on a graph. For instance, when $x = 1, f(x) = -4$. The point $(1, -4)$ would lie one step away from the graph's vertical axis and four steps below the horizontal, or $x$, axis. The resulting curve is a parabola that happens to cross the $x$ axis twice, at $x = 2$ and $x = -3$.

Solving the equation $x^2 - 5 = 0$ is a little trickier. There are two answers: plus or minus the square root of 5 ($\pm\sqrt{5}$). But what is the numerical value of $\sqrt{5}$? In other words, exactly where does the curve represented by this equation intersect the $x$ axis?

Newton's method provides one way of finding where an equation crosses the $x$ axis. The procedure begins with a guess. In the case of $\sqrt{5}$, the solver may start with 2 as a trial value for $x$. Applying the formula that lies at the heart of Newton's method produces a new, improved estimate of the root: 2.25. Now the process is repeated with the new number, and the result, 2.236, is an even better estimate. This iterative procedure continues until the solver is satisfied with the answer's accuracy.

To picture what is happening requires a look at the graph of a typical function — a smooth, undulating curve that happens to cross the $x$ axis at some point (*see Figure 6.9*). The first guess locates a point on the curve. Newton's method defines a tangent, the unique line that passes through that point and represents the slope of the curve at that point. Extending the tangent until it crosses the $x$ axis provides a new value for $x$, which is presumably a better estimate of

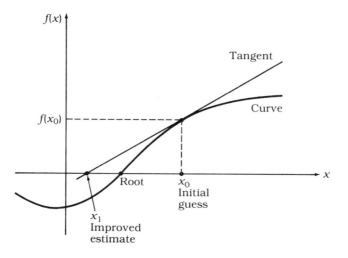

**FIGURE 6.9**    Newton's method can be used to find a root, or where a function crosses the *x* axis.

where the original curve crosses the *x* axis. The process can then be repeated by substituting this new value of *x* for the original guess, and so on.

Such a procedure can be converted easily into a reasonably efficient computer algorithm for finding one or more of the roots of practically any polynomial equation. The problem with using the method, however, turns out to be one of selecting an appropriate starting point or making the right initial guess. Not all choices of starting points converge quickly on a specific root.

The potential problems become much more serious when Newton's method is applied to functions of a complex variable. Equations involving complex numbers bring in two dimensions: each complex variable, *z*, can be thought of as having two components; whereas real numbers can be represented as points on a line, complex numbers must be located on a plane.

When complex numbers were invented centuries ago, no one could think of any practical uses for them. Now, they regularly show up in methods for solving differential equations and in other applications of calculus. They also play an important role in describing physical phenomena such as electromagnetism and the properties of electrical circuits. Complex analysis has permeated engineering.

A polynomial equation such as $z^4 - 1 = 0$ has as many roots as the highest power to which *z* is raised in the equation. In this instance, there are four complex roots. The equation $z^{17} - z^5 + 6 = 0$ has

17 solutions or roots. In terms of Newton's method, these roots act as centers of an attractive force field that draws a sequence of guesses closer and closer to one of the roots.

When Newton's method is used to find a specific root, the solver hopes that the chosen starting point leads quickly to the appropriate root. However, the method fails when the chosen point happens to fall on a boundary separating regions "controlled" by different roots. Failure may also strike when, for some starting values, the procedure gets sidetracked and, like a needle stuck on a phonograph record, ends up oscillating back and forth between two numbers, neither of which is a root.

Computer graphics provides a way to picture what is happening. For a given equation, the computer applies Newton's method for each value of $z$—perhaps a million or more different values. For each starting value, the computer determines toward which root that value converges and assigns a color to the point. Shades of the color indicate how quickly that point comes close to the root.

The result is a glowing tapestry that often features large pools of color. These basins of attraction, as they are known, are "safe" areas. Any starting point selected from a safe region, within a reasonable number of iterations, comes close to a root. The equation $z^4 - 1 = 0$ has four such basins (see Color Plate 11).

Life near or at a boundary, however, is considerably more complicated. The borders, much more than simple dividing lines, consist of elaborate swirls and whirlpools that can pull Newton's method into any one of the four roots. In these vicinities, a miniscule shift in starting point can lead to widely divergent results. And right on the convoluted boundary itself lie points that lead to no root. These points comprise the Julia set.

And there's a further complication. Magnifying a section of the boundary region doesn't simplify the border. Instead, the magnified section looks a lot like the original picture of the boundary. Further magnification merely reveals more miniature copies of the overall structure. Like a fractal, it is self-similar (see Figure 6.10).

Mathematicians are also interested in which characteristics of a function lead to particularly chaotic or unpredictable behavior. To study this problem, they pick a family of functions and vary one parameter to see what happens. One famous example, first probed by British mathematician Arthur Cayley in 1879, is the family of cubic polynomials. Cayley eventually had to give up his search because he found the answer too complicated. It took computer graphics to illustrate what was happening. The expression $z^3 - \alpha z + \alpha - 1$ represents a one-parameter family of cubic polynomials. Every polynomial in this family has a root (or zero) at $z = 1$. The location of the other two

**FIGURE 6.10**   When Newton's method is applied to the equation $z^4 - 1 = 0$, the resulting Julia set has four basins of attraction. As this closeup shows, the boundary between two such basins may be extremely complicated.

roots depends on the numerical value of the parameter $\alpha$. By varying the value of $\alpha$, which may be either a real or a complex number, mathematicians can explore how the pictures associated with each function change. That exploration involves iterating the expression $[2z^3 - (\alpha - 1)]/(3z^2 - \alpha)$.

The results are visually spectacular. At the boundaries between the basins of attraction, all three basins (and all three colors) must meet at every point *(see Figure 6.11)*. That leads to an intricate interweaving of color. Changing the parameter $\alpha$ shifts the boundaries and ruffles the basins of attraction. For certain values of the parameter, Newton's method may take a point into a never-ending oscillation between two values, interrupting its quest for a root. In that case, the point falls into an attracting cycle. In some other cases, the basin of attraction around a zero practically disappears, and no choice of points leads to the root. The zeros themselves become unstable. The disappearance of basins of attraction marks the onset of chaos.

Cubic polynomials are not the only functions worth exploring. Algebraic expressions such as $(z + \alpha)^{\alpha(z-1)}$ lead to equally startling and complicated patterns when Newton's method is applied. The resulting pictures reveal intricate basins of attractions, evidence of bifurca-

**F I G U R E  6.11**    The bull's-eye patterns indicate areas in which starting points for Newton's method converge toward one of the three roots of the equation $z^3 - 1 = 0$. The root itself falls within the small circle near the upper left-hand corner. The boundary region is a fractal because the overall pattern repeats itself on smaller and smaller scales, as seen in the teardrop shapes scattered throughout the picture.

tions, and hints of chaos. The snowmanlike Mandelbrot set appears from time to time. The Julia-set portraits of various functions show important differences as well as striking similarities.

Studying the descent of Newton's method into chaos is more than just generating pretty pictures. Ultimately, what mathematicians and scientists want is a good method for solving equations. Computer experiments and the theorems suggested by the observations are leading to a deeper understanding of Newton's method and its limitations. Mathematicians are gradually building a fence around this venerable mathematical technique for solving equations. Eventually, that fence will show where it's safe to tread and which territories to avoid.

## —— SUDDEN BURSTS AND QUICK ESCAPES ~~—

A plume of smoke rises lazily into still air. Drifting upward, it shows a smooth profile against the sky. But when the smoke column gets high enough, it begins to drift, first edging to one side, then swinging back. Soon the plume no longer has a well-defined form. It breaks up into a disorderly tangle of wisps.

The transition from a smooth to a turbulent flow can be seen not only in smoke plumes but also in water squirting from a garden hose, in air flowing over an airplane wing, and in curling ocean waves as they break into foamy whitecaps. Similar shifts from orderly to disorderly behavior show up in the populations of creatures such as gypsy moths or in the erratic chemical changes that disrupt oscillating chemical reactions. In each case, predictable behavior gradually degenerates into chaos. In well-defined stages, the systems become increasingly irregular. Simple mathematical equations, such as the logistic equation or an iterated quadratic polynomial, often fit this type of pattern.

The shift into chaos is extremely abrupt, however, in the violent contortions of flames during combustion or in pockets of agitated air that can suddenly send an airplane crashing to the ground. Such systems practically explode into chaos. Under the right circumstances, a class of simple mathematical expressions also burst into chaos. Known as *transcendental functions*, they include the exponential, sine, and cosine functions. The exponential function, represented by $e$ to some power $x$, is familiar to anyone who has dealt with compounded growth, whether in populations or in accumulated interest in a savings account at a bank. The sine and cosine functions (usually written as sin and cos), often associated with angles, come up in numerous trigonometric applications such as navigation and surveying and for describing periodic phenomena such as waves and alternating currents.

On the computer screen, iterations of these functions manifest themselves as dazzling images of iridescent dragons clawing their own tails, swirling rainbow-hued galaxies scattering vivid sparks, and geometric jets spewing colored streams into still, black basins *(see Color Plate 12)*. Moreover, by varying a single parameter, these iterated functions may burst into complete irregularity once a certain threshold is crossed.

Robert Devaney of Boston University, who has explored these functions in their complex world, has said, "Admittedly, these dynamical systems are approximate models of real physical systems, but the burst illustrated with these simple models holds the key to understanding similar phenomena in more complicated settings." Devaney's technique for studying the dynamic behavior of transcendental functions is equivalent to entering a number into the display of a scientific calculator, locating the exponential, sine, or cosine key, then repeatedly pressing the appropriate button and observing what happens to the successive numbers displayed. If the calculator is able to deal with complex numbers, it becomes possible to wander across a broad plane, not just along a narrow road.

Each iteration represents a step along a path that hops from one complex number to the next. The collection of all such points along a path constitutes an orbit. The basic goal is to understand the ultimate fate of all orbits for a given system. Depending on the value of $z$ chosen as a starting point, the orbit sometimes behaves tamely. The same answer may come up every time (a fixed point), or the answers may stay close to the original value or even return to the original value after a certain number of iterations (a periodic point). On the other hand, the answers could get steadily larger.

An orbit is considered to be stable if neighboring starting points have orbits that behave similarly; the orbits stay roughly in step. Less predictable, or chaotic, orbits have nearby points that quickly diverge. Changing the starting point $z_0$ ever so slightly may generate a vastly different orbit.

The starting points of chaotic orbits can be color-coded to indicate how quickly the points escape along their orbits to infinity. In Devaney's computer-generated maps of chaotic regions, red represents points that explode beyond a certain value in only one or two steps. The colors orange, yellow, green, blue, and violet represent successively slower rates. Black areas encompass points that upon iteration map into values that do not escape. The black areas, called basins of attraction, are stable regions. All points that are colored black, under iteration, tend toward fixed points or toward periodic points called, as we saw earlier, attractors. The colored areas represent unstable, chaotic regions. For values of $z$ in the chaotic regions, the chosen function seems to behave randomly.

The colored regions for a given complex function also give the barest outline of a Julia set, defined a little differently here than in the case of polynomial equations, where the Julia set is simply the boundary between attractors. In this case, the Julia set contains all points that seem to drive neighboring points farther and farther away. In other words, these orbits are completely chaotic: the collection of these special points corresponds to a strange repeller. The complex plane thus divides into two intricately shaped regions: basins of attraction centered on attractors and Julia sets corresponding to strange repellers.

The Julia sets that Devaney finds are often fractals as well. By examining any of the patterns closely, one finds that their features tend to replicate themselves on smaller and smaller scales. A fist bursts into fingers that each burst into smaller fingers, and so on.

A calculator experiment, using real numbers, gives a hint of how the parameter $k$ affects the exponential function $ke^x$, with $k$ always greater than zero. When $k$ is less than $\frac{1}{e}$, the function has two fixed points. Only orbits that start at real numbers greater than the largest of the fixed points escape to infinity. However, when $k$ is greater than $\frac{1}{e}$, all orbits head for infinity, no matter where they start. This behavior may be easily checked with a calculator by iterating $ke^x$ for various choices of the initial value $x$. It follows that this dynamical system has, for real numbers, two vastly different behaviors. In the complex plane, such behavior is seen as a burst into chaos.

Thus, small changes in a function can radically change the form of the resulting graphs or pictures. Multiplying the exponential by $\frac{1}{e}$ and then iterating the function over a grid of complex-numbered starting points results in a picture that shows a small, sedate jet within a large black basin. Make the constant slightly larger than $\frac{1}{e}$, and the picture changes dramatically. The Julia set explodes from a relatively small piece of the plane into two spiraling galaxies that fill the plane. Similarly dramatic changes occur when sin $z$ is multiplied by various values of a constant ranging from $1 + 0.05i$ to $1 + 0.8i$. As the imaginary part of the constant grows, the basin of attraction disappears *(see Figure 6.12)*.

Computing what happens to orbits across the whole complex plane for various values of the parameter $k$ can get quite tedious. For transcendental functions, mathematicians have found a shortcut for telling when and where a given system bursts into chaos. In the case of systems such as $ke^z$ or $k$ sin $z$, only certain "critical" orbits need be

**FIGURE 6.12**    As $c$ grows from $1 + 0.05i$ to $1 + 0.8i$, the function $c \sin z$ bursts into chaos. The dark areas represent stable regions.

followed. For the exponential function, this is the orbit of 0; for the sine function, these are the orbits starting at $\pm\frac{\pi}{2}$. Those points correspond to the location of maxima and minima in the wavelike sine function. Although the sine function has infinitely many critical points, there are only two different values. If all critical orbits go off toward infinity, then the dynamical system is completely chaotic on the whole plane. That's easy to check for $ke^z$. The starting value zero tends to infinity only if $k$ is greater than $\frac{1}{e}$.

All Devaney's results were suggested initially by computer experimentation, and many features of the behavior of transcendental functions were later established using more traditional mathematical techniques. Further explorations have located bursts in other families of complex functions. For example, a burst occurs in the $ik \cos z$ family when $k$ changes from 0.67 to a slightly larger value. Another strikes when $k$ shifts from 0 to a larger number in the family $(1 + ki) \sin z$. Rigorous mathematical proofs verify the occurrence of each burst.

Each of the many complex analytic functions known seems to have its own characteristic behavior. Just beginning their studies of iterated complex functions, mathematicians look forward to carrying their explorations across whole classes of functions, searching for common elements in their behavior. They also see value in extending the study of iterated functions into higher dimensions.

# 7

## LIFE
## STORIES

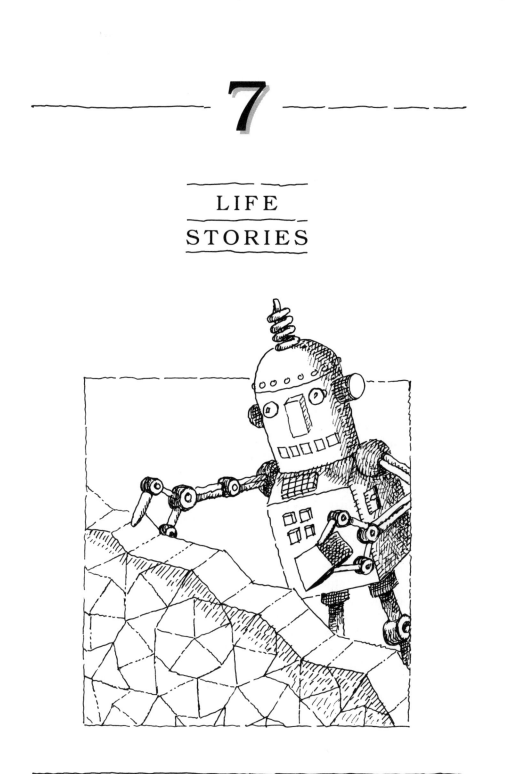

Imagine an immense checkerboard stretching as far as the eye can see. Most of the checkerboard's squares, or cells, are empty; a few are occupied by strange beings — creatures very sensitive to their immediate neighbors. Their individual fates teeter on numbers. Too many neighbors means death by overcrowding; too few, death by loneliness. A cosy trio of neighbors leads to a birth; a pair of neighbors, to comfortable stability.

At each time step, this cellular universe shuffles itself. Births and deaths change old patterns into new arrangements. The patterns evolve — sometimes into a static array that simply marks time, sometimes into a sequence of patterns repeated again and again, sometimes into a chain of arrangements that propagates throughout the checkerboard universe.

## THE GAME OF LIFE

This checkered scenario is one way of visualizing a remarkable mathematical game called "Life." Invented in 1970 by British mathematician John Conway, the game vividly demonstrates how a set of simple rules can lead to a complex world displaying a rich assortment of interesting events. Conway's aim was to create a cellular game based on the simplest possible set of rules that would still make the game unpredictable. Moreover, he wanted the rules to be complete enough so that once started, the game could play itself. Growth and change would occur in jumps, one step inexorably leading to the next. The result would be a little universe founded on logic, in which everything would be predestined. Yet there would be no obvious way for a spectator or player to determine the fate of future generations except by letting the game play itself out.

To find appropriate rules, Conway and his students investigated hundreds of possibilities. They did thousands of calculations, looking at innumerable special cases to expose hidden patterns and underlying structures. They tried triangular, square, and hexagonal lattices, covering acres of paper. They shuffled poker chips, coins, shells, and stones in search of a viable balance between life and death.

The game that they came up with is played on an infinite grid of square cells. Each cell is surrounded by eight neighbors, four along its sides and four at its corners. If a cell has exactly two neighboring cells that are occupied or alive, nothing happens. A cell that is alive stays alive, and one that is empty stays empty. Three living neighbors around an empty cell leads to tricellular mating. A birth takes place, filling the empty cell in the next generation. In such a neighborhood,

**176**

a cell already alive continues to live. But any occupied cell surrounded by four or more living cells leads to death, and the cell is emptied. Unhappily, death also occurs if none or only one of an isolated living cell's neighbors is alive *(see Figure 7.1)*.

The game's key element is that birth and death occur at the same time, not consecutively. The rules are applied to a given pattern by checking each square, one by one, and marking it as a survivor, a victim, or a newborn. All changes occur in one jump, and a new generation appears.

The game was first introduced to the public in October 1970, in Martin Gardner's "Mathematical Games" column in *Scientific American*. It aroused tremendous interest, and for many people, the game became an addictive passion. Because it was relatively easy to implement as a computer program, it also became a favorite computer exercise. All kinds of people—students and professors, amateurs and professionals—spent years of computer time following the evolution of countless starting patterns.

"Life" fanatics gleefully pursued elusive patterns and searched for unusual types of behavior. Many different forms evolved—with evocative names such as pipe, horse, snake, honeycomb cell, ship, loaf, frog, danger signal, glider, beacon, powder keg, spaceship, toad, pinwheel, and gun—and were cataloged. Some of these arrangements vegetated in a single contented state; others pulsated, switching from one configuration to another and then back again *(see Figure 7.2)*. The possibilities were endless. The game presented a variety of entertaining mathematical puzzles. For instance, are there patterns that can have no predecessor? Several such "Garden of Eden" arrays were eventually found *(see Figure 7.3)*.

Other investigations revealed that while a given pattern leads to only one sequel pattern, that uniqueness isn't present going back-

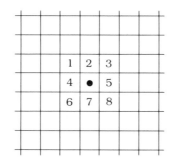

**FIGURE 7.1**   A cell's eight nearest neighbors have a strong influence on its behavior.

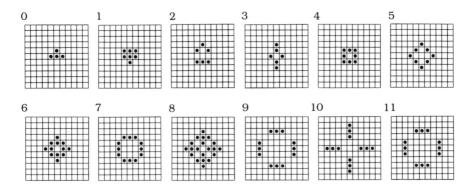

**FIGURE 7.2** One simple pattern evolves over time into a sequence that alternates between two different forms.

ward in time. A given configuration could have several possible predecessors *(see Figure 7.4)*. Thus, it could have a number of different pasts but only one future. The former property made it difficult for a viewer, glued to a computer screen, to backtrack if a particularly interesting pattern made a fleeting appearance during the course of a run. There was no simple way to program a computer to go backward in "Life."

The computer also brought animation to the game. A rapidly computed sequence of generations could produce pulsating shapes, flickering forms, creeping cancerous growth, lingering death, showers of fragments, and scattered geometric forms engaged in a chaotic dance. More recently, enthusiasts have adapted Conway's game for various surfaces other than an infinite plane. Players can

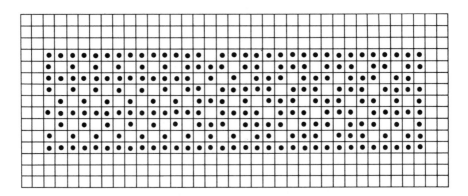

**FIGURE 7.3** A Garden of Eden array found by Roger Banks.

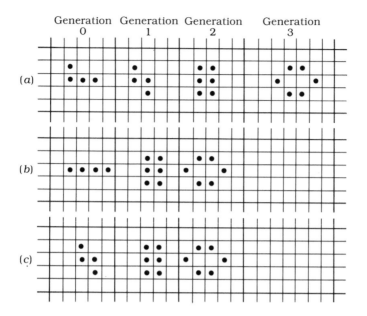

**FIGURE 7.4**    Different starting states can lead to identical vegetating states.

now follow the game on the surface of a cylinder, a torus, or even a Möbius strip. They can also pursue the creatures of "Life" on structures in three and higher dimensions.

The attraction of Conway's original game, and the chief reason for its popularity, is that although it is completely predictable on a cell-by-cell basis, the large-scale evolution of patterns defies intuition. Will a pattern grow without limit? Will it settle into a single stable object? Will it send off a shipload of colonists? Conway managed to balance the system's competing tendencies for growth and death so precariously that "Life" is always full of surprises.

## ———————— GOING WITH THE FLOW ———————

Conway's game of "Life" is an example of a cellular automaton: the mathematical equivalent of an array of simple robots programmed to do only certain tasks. The concept was introduced in the 1950s by John von Neumann and Stanislaw Ulam, who succeeded in proving that an abstract pattern could create a copy of itself by mechanically following a set of fixed rules. Their success demonstrated that self-reproduction isn't a property only of living things. Self-reproduction is also possible in the context of physics and logic. That result conjures up a surreal image of endless assembly lines of robots building robots designed for building robots.

In von Neumann's and Ulam's scheme, each cell represents an automaton, or machine, that can take on a finite number of states. For instance, a two-setting machine could be either on or off; a five-state machine would have five different settings. The choice of machine type would depend on the application. For his self-reproducing automaton, von Neumann needed 29 states. But the fate of each of the 200,000 or so cells in his self-reproducing pattern was influenced only by its own history and by the states of its four nearest neighbors.

Interesting as mathematical curiosities and games, cellular automata are now beginning to gain importance as simple models of physical phenomena, ranging from the turbulence a protruding stick generates as it interrupts the smooth flow of water to the spread of an epidemic when a contagious disease claims one victim after another. The idea is to find a way to describe a physical process compactly so that the system's main features can be followed and its future behavior predicted without having to duplicate the system in all its detail.

Traditionally, theorists have used explicit mathematical equations to represent physical processes. Physical laws are often expressed in the form of partial differential equations, which specify how variables change over time. Solving the equations produces the required prediction of a system's behavior. For example, Newton's laws of motion, embodied in a set of differential equations, provide good estimates of where the planet Jupiter will be several years from now, or where a thrown baseball is likely to land.

However, many equations, such as those describing fluid flow, are difficult or even impossible to solve exactly. To surmount this obstacle, applied mathematicians over the centuries have developed various procedures and algorithms for finding approximate solutions. The rapidly growing use of computers, which rely on approximation methods to solve equations, has focused even more attention on these methods. Recent improvements have led to efficient computer programs for solving such engineering problems as designing aircraft. Approximation methods have also lead to puzzling new concepts such as chaos, as we saw in Chapter 6.

Stephen Wolfram of the University of Illinois has developed a scheme for using cellular automata in place of computer approximations of partial differential equations to represent physical phenomena. Wolfram argues that his scheme is better suited to modern digital computers than the various methods used to approximate partial differential equations on a computer.

In a sense, Wolfram is modifying the way a theory is constructed, so that he can take the best possible advantage of an available tool. That's like changing a recipe to suit the kitchen implements on hand. The advent of massively parallel processing computers, in which networks of simple processors allow computations to be divided up so that each processor performs the same operations on different pieces of data at the same time, makes the idea even more attractive.

In 1984, Wolfram explained his approach in an article in the journal *Nature*: "Cellular automata are examples of mathematical systems constructed from many identical components, each simple, but together capable of complex behaviour. From their analysis, one may, on the one hand, develop specific models for particular systems, and, on the other hand, hope to abstract general principles applicable to a wide variety of complex systems."

Wolfram's idea, like that of von Neumann and Ulam, is to start with a lattice of sites (like the squares on a checkerboard or the hexagonal cells in a honeycomb), with each site carrying a discrete value chosen from a small set of possibilities. At every time step, these cell values are updated according to logical rules that depend on the values of neighboring sites.

The simplest types of cellular automata are narrowly confined to one dimension. A typical computer experiment starts with a line of, say, 600 sites. Each site has a value of either 1 or 0. Placing an initial pattern on the line, computing successive generations, and displaying them as black (1) or white (0) squares in sequence on a computer screen show how the pattern evolves. Like an unfurling banner, the pattern's successive rows, marking each new generation, march down the computer screen.

What happens depends on the particular rules employed. For example, the next state of a given cell may be determined by the sum of the states in the cell's neighborhood. By one such rule,

| Neighborhood sum | 5 4 3 2 1 0 |
|---|---|
| Cell's next state | 0 1 0 1 0 0 |

In this case, a neighborhood consists of a cell and two sites to the right and two sites to the left of the given cell. If each of the five cells has a value of 1, then the total is 5. According to the table, the central cell's value in the next generation will be 0. The same procedure is applied to each cell in the string.

Altogether, with two states and five-site neighborhoods, there are 64 different ways to assign values to cells in a succeeding generation. Wolfram has explored all these possibilities and many others. His empirical study suggests that patterns evolving from a simple "seed," consisting initially of just a few nonzero sites, fall roughly into four categories *(see Figure 7.5)*. Some rules don't lead to anything: the starting patterns simply die out. For other rules, the seed patterns evolve to a fixed, finite size. A third group of rules generates patterns that keep growing at a fixed rate. These patterns are often self-similar, and many have a fractal dimension of 1.59. Finally, some rules generate patterns that grow and contract quite irregularly and unpredictably.

Similar studies can be done on patterns initially consisting of, say, 100 cells, each one randomly assigned a value of 1 or 0. Again, the patterns produced by various rules tend to fall into four classes, reminiscent of the mathematical behavior of nonlinear differential equations or iterated functions under various conditions *(see Figure 7.6 and Color Plate 13)*.

About one-quarter of the rules that Wolfram studied lead to patterns that degenerate after a finite number of generations into a single, homogeneous state that endlessly repeats itself. Several of these look like vertically striped carpets. Close to 16 percent of the rules lead to a number of either unvarying or simply repeating patterns. More than half of the rules generate patterns that never develop any structure. The patterns simply look chaotic. A small number of rules generate patterns that develop complex, slowly changing subpatterns, some of which are remarkably persistent.

In general, different initial states with a particular cellular-automaton rule yield patterns that differ in detail but are similar in form and statistical properties. Different cellular-automaton rules yield very different patterns.

Wolfram has also discovered a novel, efficient way of generating random numbers. Some rules produce patterns that look completely random. In at least one case, because it satisfies several important mathematical tests for randomness, the central column from such a pattern can be used to specify a string of random ones and zeros *(see*

**FIGURE 7.5** Three of four classes of patterns generated by the evolution of one-dimensional cellular automata from simple "seeds."

*Figure 7.7).* In a sense, this particular cellular automaton hides, or "encrypts," the original data. Given only the output sequence after many generations, it would be very difficult to deduce the original seed.

A cellular automaton that appears to generate random numbers mimics the kind of behavior shown by many mathematical quantities. For example, it's easy to write down an equation that specifies the value of $\pi$, the ratio of a circle's circumference to its diameter; that equation can be used to compute $\pi$ to as many digits as desired. Yet, once the computation is done, the sequence of digits seems random, for all practical purposes. Tests on the 100 million or so known digits of $\pi$ have so far failed to turn up any evidence of an underlying pattern in the digits. Given only the digits of $\pi$, there would be no way to reconstruct the equation that defined and generated the number in the first place.

When using a cellular automaton as a model for a physical system, the challenge is to capture the key features of the phenomenon in as simple a model as possible. So far, Wolfram has built his cellular-automaton models mainly by trial and error, testing various grids and sets of rules until he finds a cellular automaton that produces the kind of large-scale behavior he wants to see.

In Wolfram's model of a flowing fluid, particles move in discrete steps along the links of a hexagonal lattice, bouncing from cell to cell. Each link supports one particle. Particles collide and scatter according to simple, logical rules, which ensure that the total number of particles doesn't change and that total momentum carried by the particles is conserved. To satisfy the second constraint, Wolfram keeps the total number of particles roughly the same in each direction on the lattice *(see Figure 7.8).*

Computer simulations of 10 million particles bouncing around in a grid show what happens. When the particle motions are averaged over large regions, large-scale flow patterns appear. For example, a cylinder (represented in the two-dimensional grid as a circle) moving through a fluid at rest, sheds the characteristic, miniature whirlpools seen in corresponding water-tank experiments *(see Figure 7.9).*

A similar approach can be used to model other physical processes, such as diffusion. In this case, particles bunch in a regular array at the center of a grid. Appropriate rules allow the particles to spread out into an apparently random arrangement. Wolfram's computer simulations show that on the average, the number of particles in a given region after a certain time interval matches the predictions of the equations commonly used to represent diffusion.

Another striking application of cellular automata involves models of the kind of branching crystal growth that leads to snowflakes and

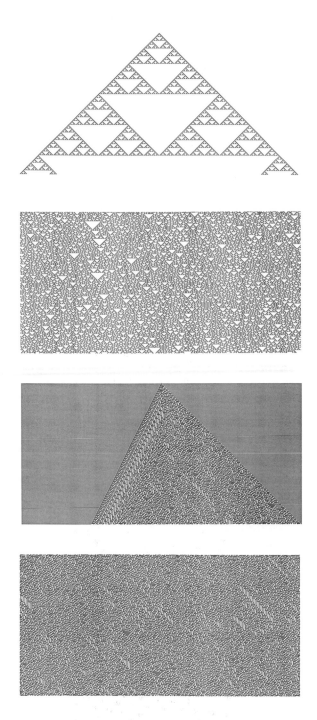

**F I G U R E   7.6**    Patterns generated by four different rules. In each pair of patterns, the upper one is obtained starting with a single nonzero site; the lower indicates what happens when a randomly chosen set of initial values is used.

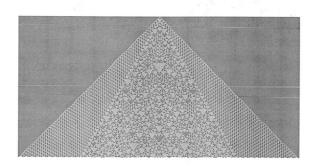

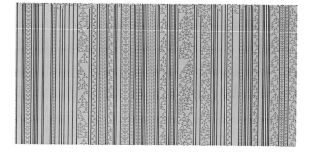

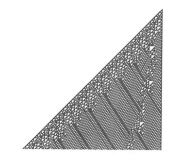

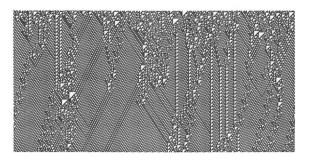

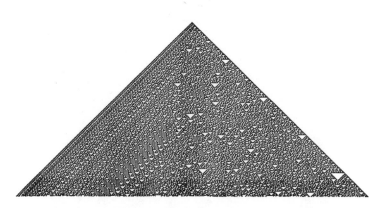

**FIGURE 7.7** A cellular automaton generates an apparently random string of ones and zeros in the center column.

other dendritic crystal forms. Governed by an appropriate two-dimensional rule, which accounts for the inhibition of growth at newly formed sites, a small seed can grow into a fractal form that resembles a snowflake.

One simple way to grow a snowflake on a computer, developed by researcher Norm Packard, is to start with a small ice hexagon at the center of a hexagonal grid. A blank hexagon turns to ice if there is an odd number of ice hexagons surrounding it. No ice hexagon forms if the cell has an even number of ice hexagons as nearest neighbors. More complicated rules lead to more intricate snowflake patterns *(see Color Plate 14)*.

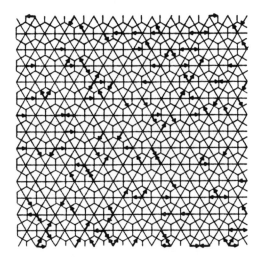

**FIGURE 7.8** In this cellular-automaton fluid model, each arrow represents a discrete particle on a link of the hexagonal grid.

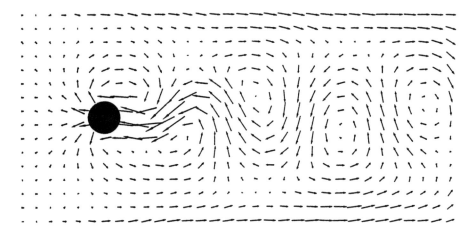

**FIGURE 7.9** A fluid-flow pattern is obtained from a simple two-dimensional cellular automaton.

The advantage of cellular-automata models is their simplicity. They mimic natural systems whose basic parts are each very simple but whose overall behavior can be very complex. What still isn't clear is how useful these models are for predicting the large-scale behavior of physical systems. There's even a possibility that some natural systems, under certain conditions, just can't be modeled by any equation or cellular automaton. In that case, using any available model, it would take so long to compute an answer in order to make a prediction that it would be easier to watch the system itself to see what it does. Prediction becomes impossible. As Wolfram notes, "Much has been discovered about the nature of the components in physical and biological systems; little is known about the mechanisms by which these components act together to give the overall complexity observed. What is now needed is a general mathematical theory to describe the nature and generation of complexity."

## FOREST FIRES, BARNACLES, AND TRICKLING OIL

Introducing an element of chance into the rules governing cellular automata brings a new dimension to studies of complex systems. Such mathematical models seem to mimic a wide range of rather unpleasant occurrences, from the spread of disease and forest fires to the growth of crab grass. These "interacting particle systems," as mathematicians prefer to call them, simulate the irregular, random spread of infestations as they jump from one victim to the next. Depending on the specific rules, all kinds of scenarios are possible.

The study of interacting particle systems is a relatively new activity for mathematicians. It began in the late 1960s as a branch of probability theory. During the two decades since then, this area has grown and developed rapidly, establishing unexpected connections with a number of other fields. The first examples of interacting particle systems were suggested by research in statistical mechanics. Physicists wanted to understand how a collection of wandering or randomly scattered particles can suddenly organize itself, as happens when a liquid solidifies or a material is magnetized.

The mathematical models developed to simulate such phase transitions proved to be a rich source of inspiration. The idea was to look at what happens to particles scattered across a grid when each particle is allowed to interact with its neighbors, according to certain rules. It became clear that models with a very similar mathematical structure could also be useful for studying neural networks, tumor growth, ecological change, and the spread of infections. Moreover, the dynamical behavior of such models suggested intriguing mathematical questions. Mathematicians became interested in how certain models evolve over time and began searching for unusual types of behavior.

Cornell University's Rick Durrett, one of the leaders in this new area of mathematics, says, "Mostly, it's a mathematician's game of seeing what happens." The mathematician defines the rules, sets up the game board, and lets the game play itself out.

One example of an interacting particle system is a model that resembles the inexorable spread of a raging forest fire, as it fans out in a blazing ring, consuming fresh timber at its outskirts and leaving a burnt residue within. The rules for a mathematician's forest fire are simple. The playing field is a checkerboard grid on which each cell represents a tree. The fire begins as a single, marked cell or a small cluster of cells at the grid's center. At each time step, a burning cell has a certain probability of spreading the fire to its four nearest neighbors, unless those neighbors have already been burnt.

The same rules also lead to a rough model for the spread of an infectious disease such as measles. In this case, each cell represents an individual who is either healthy, ill, or immune.

The simplest set of rules mathematically worth investigating makes the unrealistic assumption that the fire (or infection) in a given cell lasts for only one unit of time. At each step, the toss of a coin or a similar randomizing procedure decides whether a certain burning cell spreads its flames to each of its neighbors. Hence, at any given time, a computer screen would show three types of cells: those that are burn-ing, those already burnt up, and those that have so far escaped un-scathed. The burning cells usually form an irregular, broken ring that gradually expands as time goes on.

Mathematicians, particularly probabalists, are interested in how the process depends on the probability of transmission from one cell to another. They find that when a cell has only 4 chances in 10 (a probability of .4) of passing on the fire to a given neighbor, then the fire eventually dies out. If it has 6 chances in 10 (a probability of .6) of doing the deed, then the fire continues to spread. The critical proba-bility that establishes the dividing line between these two types of behavior turns out to be .5.

The shape of the ring also seems to depend on the transmission probability. At a probability of .5, the ring's outer edge looks like a fractal, having an incredibly convoluted boundary that when magni-fied, instead of looking smoother, appears to be equally complicated on every scale. Theorists have no idea why a fractal should appear at this critical value. By the time the probability reaches .6, the ring is smoothed out into a circle. At even higher probabilities, the ring takes on the shape of a square (see Figure 7.10).

Oriented percolation, another example of an interacting particle system, resembles the downward trickle of a surface pool of oil through an underlying bed of sand or clay. Fingers of oil penetrate this layer if the ground is porous enough. The trick is to find the critical probability at which a sufficient number of air spaces are present so that an open pathway exists. Because no one knows pre-cisely what this critical probability is, theorists must resort to com-puter simulations to get a feel for what the process looks like.

The oriented percolation model can be pictured as a gigantic diamond-shaped grid of connected pipes (see Figure 7.11). A valve in each section of pipe in the grid may be either open or closed. At the top of the network is a reservoir of fluid. If all valves are open, fluid will flow down through the network of pipes. If all valves are closed, no flow occurs.

What if some valves are open and others closed? Clearly, no fluid will flow until enough valves are open. By studying the results when the probability of a particular valve being open lies somewhere be-tween 1 (all valves open) and 0 (all valves closed), mathematicians can determine the critical probability at which flow is first established.

Computer experiments show that when the open-valve probabil-ity is .55, the process dies out. This means that on the average, when 55 out of 100 valves are open, fluid gets into the pipes but doesn't get very far before all paths for fluid flow are blocked. As the probability

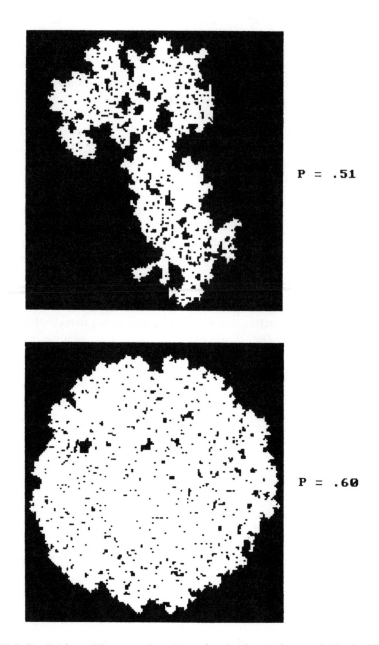

P = .51

P = .60

**FIGURE 7.10**    The growth pattern for the forest-fire model looks like a fractal when the probability of transmission is near .50 but is roughly circular at a higher probability.

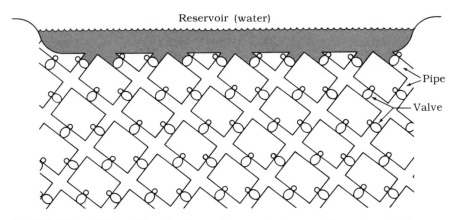

Reservoir (water)

Pipe

Valve

**FIGURE 7.11** How far the water penetrates into this network of pipes depends on which valves are open.

that a valve is open increases, the fluid penetrates deeper. Percolation — long-distance flow — is established when the probability of an open valve gets close to .645 *(see Figure 7.12 and Color Plate 15)*. However, a mathematical proof that this number must be the critical value has been elusive. The best that anyone has done is prove that the lower bound on the critical probability is .6298 and the upper limit .84.

A similar, two-dimensional model can be set up for the spread of a plant species or the propagation of a population of immobile animals such as barnacles or mussels. This particular model is named for British mathematician Daniel Richardson of the Polytechnic of the South Bank in London, who first suggested the model in a 1973 paper.

Partly because there is no provision for death, the model's pattern always grows to cover the entire plane, leaving only a few holes near its fringes, while growth occurs. Theorists are interested in the times at which a growing shape exceeds a certain boundary and in how this time depends on the probability of an occupied cell sending an offspring (or root) to an adjacent, unoccupied space *(see Figure 7.13)*.

The examples illustrate just a few of the infinite number of possible models that could be investigated. Even in a problem as simple as the forest fire model, the number of possibilities is striking: the rules could be changed to involve a different number of neighbors; the burning stage could last for more than one unit of time; the grid itself could be hexagonal or triangular instead of square. The possibilities seem limited only by the researcher's imagination.

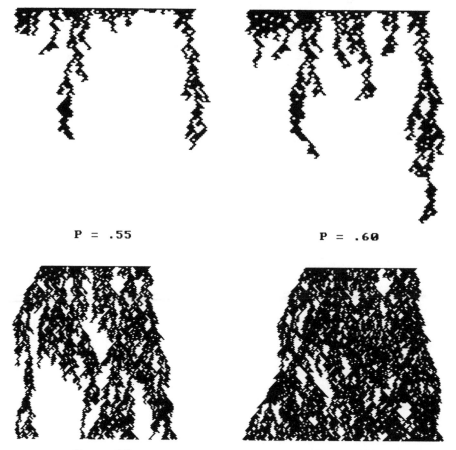

P = .55    P = .60

P = .65    P = .70

**F I G U R E  7.12**    Changing the probability that a valve is open changes how deeply a fluid penetrates the matrix.

How do mathematicians decide which models are worth studying? Some models are suggested by studies in other fields, such as physics and biology. Others show some type of novel behavior. In general, most models do nothing interesting. Only a few do something special, and these are the ones that mathematicians seek out and study. And there's another barrier. Many candidate models are too complicated to be analyzed mathematically. Mathematicians keep trying rules until they find a system they can analyze.

Both the emphasis on diversity and the difficulty mathematicians have in rigorously proving general statements about these models lead to what seems a curiously fragmented type of mathematics. The unity of the subject doesn't come from general theorems, as in other fields of mathematics. Instead, it comes from the same techniques being used to analyze a variety of models. Mathematicians find it very difficult to come up with statements that describe all interacting particle system models.

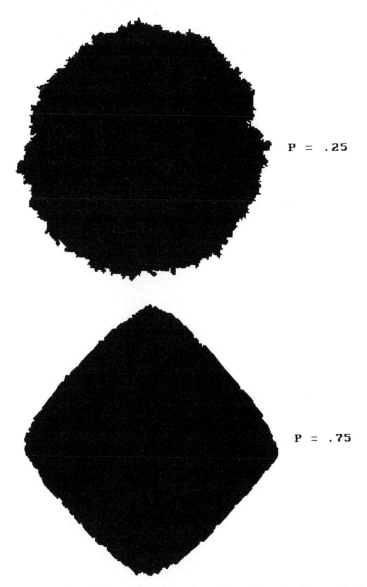

P = .25

P = .75

**FIGURE 7.13**   The probability of sending offspring to a neighboring cell affects the geometric shape of the pattern.

The way various interacting particle systems behave in different dimensions is a topic that has produced some interesting results. Such models need not be restricted to planar cells or even to three-dimensional blocks. Although difficult to visualize, these models can be studied in any dimension. Mathematicians have found that the behavior of models changes in different dimensions. Sometimes, they find a critical dimension, which means that at lower dimensions one thing happens and at higher dimensions something else happens.

In fact, the behavior of most models seems to get simpler when the dimension gets larger. Because higher dimensions have so much room, the interactions between particles get weaker. Particles tend to be much farther apart. Frequently, higher-dimensional versions of various processes behave almost as if there were no interactions. They begin to look a lot like processes such as random walks, traditionally studied in probability theory and made up of independent steps.

So far, most interacting particle models are not complicated enough to be taken seriously as stand-ins for natural processes. But with the increasing use and greater sophistication of computer simulations, it's possible that interacting particle systems may eventually find uses in fields such as ecology and geology. "Many of these problems can be explained in a few minutes to a person with no prior experience in the area," says Durrett. "That's . . . one of the charms of the area. It's not something that you have to spend three years in graduate school to appreciate. And it's easy to play around with these things on a computer."

## MATTERS OF STATE

One of the most provocative questions in mathematics is whether a mechanical way exists to determine the truth of any given mathematical statement. Is some kind of universal machine, following a strict procedure, capable of judging mathematical truth?

With one qualification, the surprising answer is yes. Such an all-purpose machine not only is possible, but also can do anything that a human mathematician or the most powerful computer conceivable in theory or in practice can do. However, it can't answer mathematical questions that human mathematicians themselves can't answer using a mechanical procedure. For instance, it can't determine the digits of a nonrepeating decimal, such as $\pi$, because any mechanical method would require an infinite number of steps.

Mathematician Alan M. Turing was one of the first to propose the idea of a universal mathematics machine. Turing and Emil Post independently proved that determining the decidability of mathematical propositions is equivalent to asking what sorts of sequences of a finite number of symbols can be recognized by an abstract machine with a finite set of instructions. Such a mechanism is now known as a Turing machine.

A Turing machine can be pictured as a black box capable of reading, printing, and erasing symbols on a single, long tape or strip of paper divided by lines into square cells, or boxes. Each cell is blank or contains one symbol from a finite alphabet of symbols.

The Turing machine scans the tape, one cell at a time, usually beginning at the cell furthest to the left that contains a symbol. The machine can leave the beginning symbol unchanged, erase it, or print another in its place. If in reading the tape the machine later encounters a blank cell, then the machine has the choice of leaving the cell empty or entering a symbol. After it performs its assigned task on a given cell, the machine stops or else moves one cell to the left or right.

What the machine does to a cell and which way it moves afterward depends on the state of the machine at that instant. Like a state of mind, the machine's internal configuration establishes the environment in which a decision is made. Turing machines are restricted to a finite number of states.

An *action table* stipulates what a machine will do for each possible combination of symbol and state. The first part of the instruction specifies what the machine should write, if anything, depending on which symbol the machine sees. The second part specifies whether the machine is to shift one frame to the left or to the right along the tape. The third part determines whether the machine stays in the same state or shifts to another state, which usually has a different set of instructions.

Suppose a Turing machine must add two integers. There are numerous different ways in which this can be done, depending on how many symbols and states are allowed. One of the simplest possibilities is to represent an integer by a string of asterisks. Thus, *** would be 3, and **** would be 4. To add 3 and 4, *** and **** are first printed on a tape, with a blank space between the two strings. The machine then fills in the blank cell by printing a * and goes to the end of the string of asterisks and erases the last one in the row. What's left is the required answer: a set of 7 asterisks.

An action table instructs the machine how to perform the addition. The table's first column gives the machine's possible mental states, and the first row lists all the symbols being used. In this

|  | *Symbol* | |
|---|---|---|
|  | * | Blank |
| State 0 | right, state 0 | print * , right, state 1 |
| State 1 | right, state 1 | left, state 2 |
| State 2 | erase, stop | — |

example, there are only two symbols: an asterisk * and a blank. Each combination of symbol and state specifies what, if anything, needs to be done to a cell, in which direction to move after the action, and the state of the machine, that is, which set of instructions it will follow for its next move.

The machine begins in state 0 and scans the * farthest to the left on the tape. According to the instructions for state 0 and symbol *, the machine leaves the symbol as it is, shifts one step to the right, and remains in state 0. It encounters another *, and the process is repeated. Finally, it reaches the blank cell. From the table, the machine knows it must print a *, then move one space to the right. This time, however, it shifts into a new state. Now the machine stays in state 1 until, step by step, it reaches the first blank at the end of the string. This time, it backs up one space and shifts into state 2. It erases the cell's asterisk and stops. The addition is complete *(see Figure 7.14)*.

The same action table can generate the sum of any two whole numbers, no matter what their size. But adding two numbers such as 49,985 and 51,664, by itself, would require a tape with at least 100,000 cells. To be capable of adding any two numbers, the tape would have to be infinitely long. In fact, a universal Turing machine capable of any mathematical operation, must have an infinitely long tape. An ordinary computer, with a limited amount of memory, lacks this property.

Similar tables can be worked out for subtraction and for practically any other mathematical operation. The sole condition is that the number of states and symbols listed in such a table is finite, which ensures that a routine, mechanical process can do the job. Often, several different tables can be used to perform a certain operation.

The world of Turing machines has countless nooks and crannies worth exploring. One such corner is occupied by a particular group of Turing machines dubbed busy beavers. These machines, following an action table with a strictly limited number of states, must print as many symbols (say, the digit 1) as possible on an initially blank tape before they grind to a halt. The champion machine for a given number of states—the busy beaver—is the one that prints the maximum number of ones.

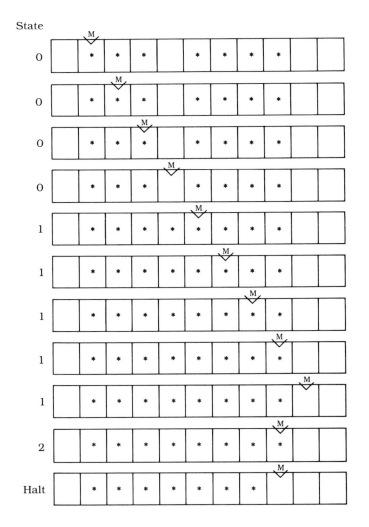

**F I G U R E  7.14**     At each step, a Turing machine may move one space to the left or right. By following a simple set of rules, the machine can add two numbers, starting with separate groups of three and four asterisks and ending up with one group of seven asterisks.

Mathematicians already know that a three-state busy beaver writes 6 ones before it comes to a halt *(see Figure 7.15)*. The four-state busy beaver stops after writing 13 ones. The answer for a five-state busy beaver isn't yet known, but one Turing machine that prints 1,915 ones has been found.

The discovery in 1984 of this particular five-state automaton was the work of an amateur mathematician, George Uhing. Uhing built a

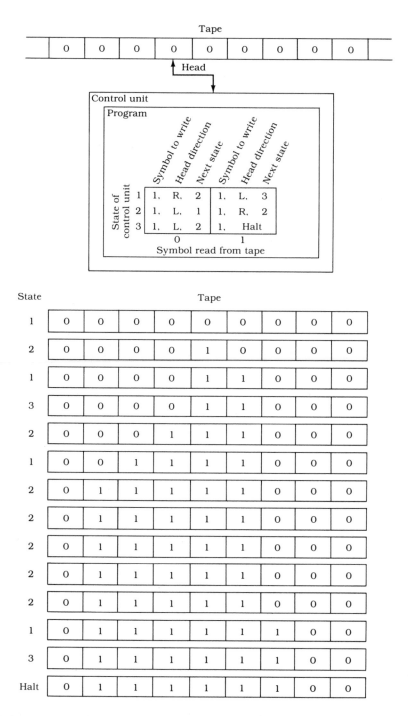

**FIGURE 7.15** A three-state, busy-beaver Turing machine, starting with a blank tape, writes a string of 6 ones before it comes to a halt. A simple program tells the machine what to do at each step, depending on whether the machine reads a one or zero.

small computer that automatically tested, one after another, a selection of the 64,403,380,965,376 different possible five-state, two-symbol (1 and 0) Turing machines. Many of the candidates are easy to eliminate because their action tables lead them into infinite loops, and the machines run on forever. Other halfhearted contenders stop almost immediately.

After letting his computer run for about 3 weeks while it sifted through several million possibilities, Uhing found the five-state Turing machine that prints out 1,915 ones, making more than 2 million moves to generate the string. The string itself consists of alternating ones and zeros, except for the last seven spaces, which consist of "110 1111."

Whether this particular machine is the long-sought ultimate busy beaver isn't known. Other five-state machines, not yet identified, may print out even more ones. However, the fact that a five-state Turing machine can print at least 1,915 ones is itself significant. Uhing's result implies that the behavior of even a simple digital machine can quickly get out of hand, much sooner than mathematicians had expected. Translated into mathematical terms, the large jump evident in the output from a four-state to a five-state busy beaver means that the amount of computation, or the number of steps needed to solve particular problems, can easily outstrip the ultimate problem-solving capacity of real computers.

Moreover, determining whether a particular Turing machine stops is tantamount to proving or disproving certain mathematical conjectures. Uhing's result diminishes mathematicians' chances of distinguishing between machines that halt and machines that don't halt. The difference between a decidable and an undecidable proposition may be as subtle as the difference between a machine that runs on forever and one that goes on so long that everyone loses patience waiting for it to stop.

The most important part of a Turing machine is its action table, which is a bit like the software that instructs a computer; a Turing machine itself is somewhat like a typewriter whose characters can be put together in countless different ways. The same typewriter can produce any novel that has ever been written or ever will be written. Similarly, as Turing and Post were able to prove, a universal Turing machine can be programmed with a finite set of instructions to imitate any other special-purpose machine.

What is remarkable about Turing machines is that these simple devices embody a method of mathematical reasoning. Although Turing machines seem very limited, they can perform any mathematical operation that a human mathematician or a computer can perform. Given a large but finite amount of time, a Turing machine is capable of any computation that can be done by any modern digital computer, no matter how powerful. A supercomputer may do the job faster, but the slowpoke Turing machine eventually reaches the finish line, too. Moreover, a Turing machine doesn't have to be a machine. It can be a computer, a mathematician, a team of students, a game, or any other entity that follows an action table.

Interestingly, John Conway was able to prove that the patterns observed in "Life" can function as universal computers; in other words, the game has enough suitable patterns for the construction of a Turing machine. In fact, the circuitry of any possible computer can be built from a set of four patterns that appear in the game: guns, gliders, eaters, and blocks.

Going one step further, the "Life" universe itself can be thought of as an array of computers or a single massively parallel computer. The same rules are applied to each cell, and the "Life" universe accepts data in the form of a particular starting pattern of marked or living cells. After the rules are applied, the new pattern of living cells comprises the output data. In this way, the infinite plane on which "Life" is played supplies the answers to mathematical questions.

Wolfram's one-dimensional cellular automata also show signs that at least some of them may have the properties necessary to be Turing machines. Von Neumann's self-reproducing cellular automaton itself actually incorporates a Turing machine. By including a universal computer, which extends its reach into empty regions of the cellular plane, his machine also becomes a universal constructor, able to make copies of itself.

## —— —— —— THE FIVEFOLD WAY ——

In its simplest form, a tiling problem in mathematics is not unlike the practical task of covering a floor or wall with ceramic tiles. Tiles in the shape of equilateral triangles, squares, or hexagons are particularly easy to work with and do the job nicely (see Figure 7.16). The simplicity and symmetry of the resulting patterns are reasons why researchers interested in cellular automata tend to stick to triangular, square, and hexagonal grids for their models.

An undecorated, triangular tile with sides of equal length has a threefold symmetry. When it is rotated through one-third of a circle, or 120 degrees, nothing seems to change. In its new position, the tile looks just as it did in its old position. That isn't necessarily true of a decorated tile carrying a pattern. Three 120-degree rotations, through a total of 360 degrees, bring the tile back to its original position.

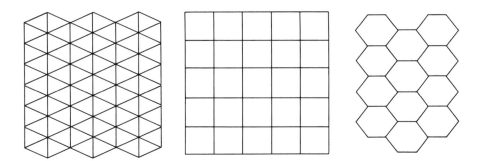

**FIGURE 7.16** Regular polygons, such as triangles, squares, and hexagons, can be put together to generate symmetric tilings.


Similarly, plain, square tiles have a fourfold symmetry, and hexagons have a sixfold symmetry.

The same symmetry considerations also apply to the patterns formed when tiles are laid down side by side. If a pattern can be rotated through, say, 90 degrees, and it appears identical to the pattern in its original orientation, then it, too, has a fourfold symmetry.

Tiles in the shape of regular pentagons, with five sides of equal length, have a fivefold symmetry. Attempts to lay such tiles out on a flat surface invariably leave embarrassing gaps in the patterns because the pentagons don't fit together to cover a plane completely. Using a pair of figures that came to be known as darts and kites, mathematical physicist Roger Penrose of Oxford University solved the pentagon tiling problem in 1974. His discovery came in the course of a search for tile shapes that can tile the plane nonperiodically but not periodically. Whereas a tiling pattern made up of, say, squares repeats itself at regular intervals, and the tiles line up in neat rows that are whole-number distances apart, the resulting pattern in the kite-and-dart case doesn't repeat itself at regular, whole-number intervals. Instead, the irrational number $(1 + \sqrt{5})/2$, also known as the golden ratio or mean, pervades the pattern.

A pair of the required shapes — a dart and a kite — can be created by cutting a rhombus, a figure resembling a skewed square or a diamond, into two pieces in a special way (*see Figure 7.17*). The rhombus itself must have internal angles of 72 and 108 degrees. Of the two diagonals that can be drawn from corner to corner across the rhombus, the longer one is divided in the golden ratio; that is, one segment ends up roughly 1.618 times longer than the other. Joining this special point on the diagonal to the two other corners divides the rhombus into two areas. One piece of the rhombus resembles an

**201**

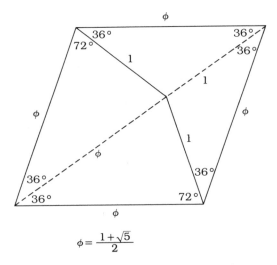

$$\phi = \frac{1+\sqrt{5}}{2}$$

**F I G U R E** 7.17    Dividing a rhombus in a special way produces a kite and a dart.

arrowhead. The other piece, whose blunter end fits snugly into the arrowhead's notch, looks like a diamond with one foreshortened end. Not surprisingly, when tiling the plane, it's possible to reassemble kites and darts into rhombuses and to arrange these forms in such a way as to create an ordinary, regularly repeating arrangement. To get a nonperiodic pattern, the tiles must be forced to fit together only in certain ways. That can be done by decorating the appropriate sides or angles of each dart and kite with colored bands or by using small, interlocking tabs like those on jigsaw-puzzle pieces *(see Figure 7.18)*. Such rules for matching embody the restrictions on how the tiles may be laid down next to one another.

Penrose, working with John Conway, proved there are infinitely many ways to cover the whole plane with kite-and-dart tiles. None of these tilings, or tessellations, is periodic. Some patterns are symmetric, but most are a mystifying blend of apparent order and unexpected deviations. The patterns seem to strive for regularity, but they always fall just short of it.

Although there are an infinite number of possible patterns, one pattern can be distinguished from another only when the tiling, spread out over an infinite plane, is viewed as a whole. An explorer, suddenly parachuted onto such a patterned plane, would have no way of telling on which tiling he had landed. No matter how much he wandered about the plane, he would always find himself in vaguely familiar surroundings, yet he would be forever lost.

Furthermore, in the tightly knit family of Penrose tilings, every finite region in any pattern sits somewhere inside every other pattern. A tourist, wandering across any Penrose tapestry, would eventually encounter, more or less randomly, every possible finite pattern into which the tiles can be formed—a veritable encyclopedia of arrangements.

Remarkably, each plane-filling tiling pattern contains exactly 1.618 . . . times as many kites as darts. Because this ratio is an irrational number, it's impossible to break down the tiling into a single unit cell containing an integral number of each kind of shape. If it existed, such a unit cell could be used as a building block to construct a periodic Penrose tiling.

To prove that his kite-and-dart tiles really do fill a plane and always produce nonperiodic patterns, Penrose showed that his matching rules are closely related to a set of inflation and deflation rules. Symmetrically bisecting each dart in any tiling, then collecting together the resulting half-darts and kites, makes kites and darts on a larger scale. That is, two half-darts and one kite make a large dart, whereas two half-darts and two kites make a large kite. These larger figures can be used just as easily to build up a complete tiling pattern, and the same process of divide and collect can be applied to them, ad infinitum. The process of inflation guarantees that kites and darts fill the plane (*see Figure 7.19*).

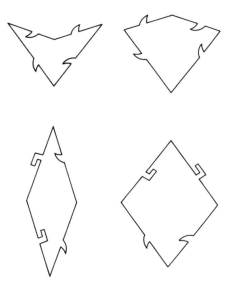

**F I G U R E   7.18**   Adding notches and projections ensures that tiles can't be put down in a periodic pattern.

**FIGURE 7.19**     Inflation rules show that kites and darts fill the plane.

Equivalently, by deflation rules, each tile shape can be divided up into smaller units of the same shape. If each of the tiles in a cluster, constructed according to the matching rules, is divided up or deflated, a new cluster is generated, containing more tiles in which the rules for matching are still obeyed. By repeating the deflation over and over, one can extend the matching rules to a tiling that consists of an infinite number of units. The deflation rule also means that a Penrose tiling's pattern is self-similar.

As it turns out, there are many different pairs of quadrilateral shapes that form a complete tiling pattern, although all are related in some way to the original kite-and-dart pair. All pairs have characteristic lengths whose ratio is the golden mean. One particularly useful set is a pair of diamond-shaped figures — one fat and one skinny *(see Figure 7.20)*. The skinny diamond has internal angles of 36 and 144 degrees; the fat one, angles of 72 and 108 degrees. Not surprisingly, 1.618 . . . times as many fat diamonds as skinny diamonds are needed for a complete tiling pattern *(see Figure 7.21 and Color Plate 16)*.

Penrose also played the three-dimensional version of his tiling game: packing space with simple blocks, such as pairs of squashed

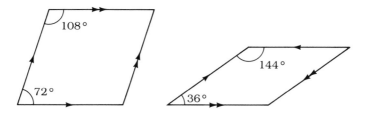

**FIGURE 7.20** A pair of rhombuses can also be used to tile a plane nonperiodically.

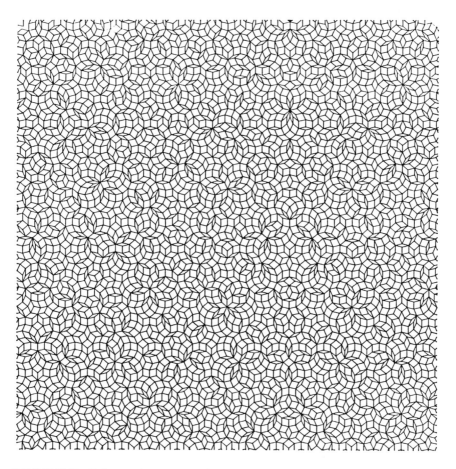

**FIGURE 7.21** A Penrose tiling using thin and fat rhombuses.

cubes (rhombohedra), that generate nonperiodic solid structures with fivefold or, in three dimensions, icosahedral symmetry. The icosahedron, which serves as a model of fivefold symmetry in three dimensions, is a solid figure consisting of 20 identical triangular faces, with 30 edges and 12 vertices. Each vertex lies on one of six fivefold symmetry axes connecting opposite vertices *(see Figure 7.22)*. Just as pentagons can't be used to tile the plane, icosahedra can't be packed together to fill space completely.

Pairs of rhombohedra can, however, be used to fill space, generating the three-dimensional analog of Penrose's kite-and-dart patterns. Again, no unit cells serve as building blocks, and matching rules lead to nonperiodic structures, where the ratio of the number of rhombohedra of each type is the golden mean.

Penrose tilings have many other intriguing features and suggest a wide range of mathematical questions, many of which haven't been answered yet: Is there a single shape that can tile a plane in a nonperiodic way but cannot tile the plane periodically? Are there pairs of tiles that tile the plane nonperiodically but are not related by the golden ratio?

One perplexing question arises out of the frustrations associated with building a Penrose pattern. It's always possible to continue laying down pieces forever, but finding the right course to take is seldom obvious. Sometimes the choice of where to put a piece is forced, and sometimes there are several alternatives. Making the wrong choice leads to a spot where no piece can legally be put. No one yet knows

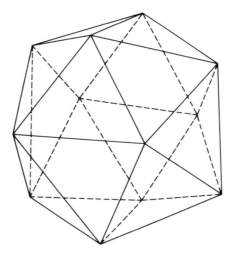

**FIGURE 7.22** An icosahedron has a fivefold symmetry in three dimensions.

whether it's theoretically possible to find a general technique for predicting which starting patterns lead to a full tiling of the plane.

Roger Penrose didn't have anything practical in mind for the remarkable tiling patterns he created when he started drawing his diamond tapestries more than a decade ago. The exercise was simply a challenging mathematical game that tickled his fancy. Nevertheless, Penrose did consider some implications of his findings for crystallography, the study of crystal shapes. Although Penrose saw that some concepts crystallographers accepted without question were not always strictly true, he couldn't easily counter the traditional rules that crystallographers had been following for more than a century. The rules rigidly maintained that nonperiodic structures are forbidden because the units of atoms that make up crystals must fall into an orderly, regular arrangement in order to fill space completely. In common salt, for instance, sodium and chloride ions sit at the corners of a cube, and these cubes stack neatly to fill out each salt crystal.

A few people did take Penrose's ideas seriously, but it took the discovery in 1984 of tiny metallic crystals composed of aluminum and manganese to alert the scientific world to Penrose's tilings. These crystals had a form as startling and unexpected as five-pointed snowflakes. X-rays and electrons reflected from the crystals showed Penrose's fivefold symmetry—an event, according to the long-standing rules of crystallography, that wasn't possible (see Figure 7.23).

The arrangement of atoms in a solid is comparable to the placement of tiles in a mosaic. In crystals, atoms or clusters of atoms appear in repeating motifs, analogous to tiles that fill in a mosaic. The repeating motifs, called unit cells, join together to form the complete crystal structure.

Crystal structures generally have a high degree of order. A microscopic observer standing inside a crystal would see row after orderly row of atoms. If the observer were to shift to another position, exactly one unit cell away in any direction, the view would be identical. Just as the position of one brick decides the positions of all other bricks in a perfectly regular wall, the position of one unit cell determines the positions of all other unit cells. Similarly, the orientation of one unit cell sets the orientations of all unit cells.

This requirement for order in a crystal seems to imply that only triangular, square, or hexagonal lattices and variations on these forms have the necessary regularity. According to the well-established suppositions of crystallography, formalized in the branch of mathematics known as group theory, only a small list of rotational symmetries is possible for crystals. For instance, a crystal is said to have threefold rotational symmetry if the crystal latticework would not change in appearance after the crystal is rotated 120 degrees

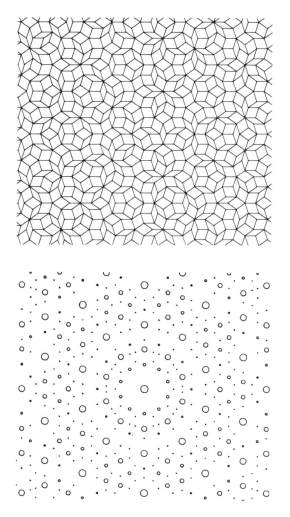

**F I G U R E  7.23**  A Penrose tiling *(top)* produces a diffraction pattern of spots *(below)* that has a fivefold symmetry.

about one of its axes. In general, crystals can have twofold, threefold, fourfold, or sixfold axes of rotational symmetry.

However, Penrose's patterns show that fivefold symmetry is possible if more than one unit cell is used and if the need for full geometrical regularity is loosened a bit. In this case, the tiles are neither randomly spaced nor made to fall in neat rows. The position of one tile determines in predictable but subtle ways where the remaining tiles fall. Patterns form, but they never quite repeat themselves. Moreover, each tile of a given shape is oriented along one of a small, discrete set of special directions. For example, Penrose tilings show a scattering of decagons — ten-sided figures that at a distance could be mistaken for circles *(see Figure 7.24)*. All these decagons have the same orientation throughout a pattern: the sides of one decagon are parallel to the corresponding sides of all other decagons.

The newly discovered crystals appear both to be highly ordered and to have a fivefold or, in three dimensions, an icosahedral symmetry. The evidence for both qualities is in the sharply defined spots seen in x-ray and electron diffraction analyses done on the crystals. The mathematical model that best accounts for these properties seems to be one based on the Penrose tilings. The idea would be to use rhombic triacontrahedra, each of which is made up of ten thin and ten fat rhombohedra, as units to fill space *(see Figure 7.25)*. These units would fit together according to matching rules similar to those that govern Penrose's tilings in two dimensions.

But materials scientists and crystallographers aren't ready to give up their notions of periodicity entirely. They can save the situation to some degree by going into a higher dimension. A simple analogy illustrates the principle involved: suppose a two-dimensional crystal is made up of atoms that sit in a regular square array, a lattice of equidistant points, which is clearly periodic *(see Figure 7.26)*. A line is drawn in a random direction passing through the lattice. If each point in the neighborhood of the line casts a shadow onto the line, the positions of the shadows will no longer be evenly spaced. Thus, an arrangement of points that has neat, evenly spaced rows in

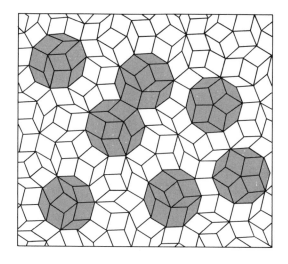

**F I G U R E  7.24**    Decagons have the same orientation throughout the pattern.

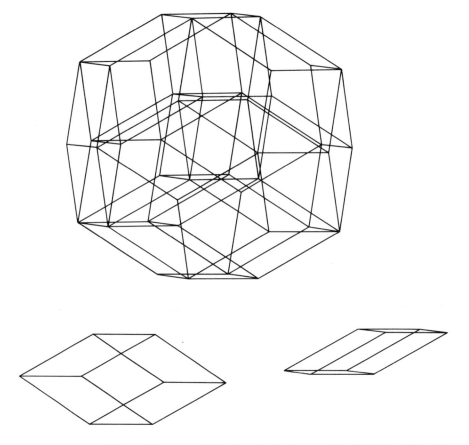

**FIGURE 7.25** Sharp and flat rhombohedra can be fitted together to produce a rhombic triacontrahedron.

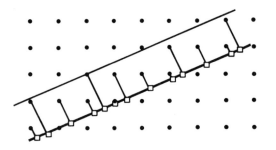

**FIGURE 7.26** A lattice of equidistant points projects into a quasiperiodic array of points along a line.

two dimensions no longer looks periodic when projected onto a one-dimensional line. In one dimension, the projected points are said to be quasiperiodic.

When a sample of the strange, new crystalline material is rotated in an electron beam, the diffraction patterns show evidence that these crystals have many fivefold as well as twofold and threefold axes. By noting the angles between pairs of these axes, it's possible to conclude that the material has the same symmetry as an icosahedron. The six icosahedral axes showing a fivefold symmetry form the basis for a six-dimensional space. In this space, the crystal would have a cubic unit cell, which means that these unit cells will stack evenly to form a regular, periodic crystal lattice. However, when the six-dimensional structure is viewed as a projection into three dimensions, the periodicity disappears. The three-dimensional projection looks quasiperiodic.

Because the new structures are highly ordered—just like normal crystals—but are quasiperiodic instead of periodic, they came to be called quasiperiodic crystals, or quasicrystals for short. Like Superman, they have a double identity. In six-dimensional space, they lead a regular, perfectly ordinary existence. In three-dimensional space, they have extraordinary properties, breaking the rules that appear to govern other crystalline materials. These new substances bring crystallography into higher dimensions and present crystallographers with a new mathematics to learn.

Of course, the full story isn't quite so simple. Some scientists still dispute the discovery of quasicrystals, claiming that the crystal samples are really made up of twins, which by themselves are regular, but together create the anomalous diffraction patterns. On the other hand, those who believe in quasicrystals debate how the atoms within the rhombohedral unit cells may be arranged. They argue the fine points of decoration—the position of each aluminum and manganese atom within the postulated unit cells.

Furthermore, there are ways to describe a quasicrystalline lattice without having to venture into six dimensions. Mathematician Branko Grünbaum, an expert on the mathematics of tilings, argues strongly in favor of *adjacencies* to describe tilings and crystal forms, an alternative to the group-theory approach traditionally used by crystallographers and other researchers. Whereas group theory emphasizes the importance of a pattern's overall rotational-symmetry properties, the adjacencies model takes a local approach, specifying how one piece is connected to its neighbors. In fact, adjacency rules generate many interesting patterns that can't be described by symmetries and group theory.

In a sense, the adjacencies way of looking at tilings isn't particularly new. For millenia, artists and craftsmen have created intricate, pleasing designs and repeating patterns on walls, in woven fabrics, on pottery without relying on the tenets of group theory to guide their work. They counted on their own empirical rules and schemes, which more often than not simply stated how one tile or shape fits with another. Once the rules and tile shapes are selected, the types of patterns that emerge are set.

Mathematicians face other challenges as well. They already know that simple matching and inflation rules for various types of two-dimensional tiles can force the creation of patterns not just with a fivefold symmetry but also with eightfold and twelvefold symmetries. Are there other possibilities? The same question applies to three-dimensional shapes for filling space. Is the icosahedral form the only one that occurs in three dimensions? Even physicists, now on the lookout for even stranger crystalline forms, may be interested in the answers.

And, what happens if the game of "Life" is played on a Penrose tiling? How should the rules be modified to keep the game interesting? The prospects are mind boggling.

# 8

## IN
## ABSTRACT
## TERRAIN

$T$ake a generous helping of proof—the essence of serious mathematics—and mix it with a dash of clever computer science. Add a sprinkle of scrambled graphs, throw in a touch of chance, and spice it with the tang of secrecy. The result is a piquant brew that adds up to a playful, foolproof method for keeping a secret. It gives a wary mathematician, caught in the sometimes turbulent world of mathematical research, an ingenious way to claim credit for being the first to find a particular proof without having to give away the slightest clue as to what the proof is. All that a rival can find out, until the proof itself is finally revealed, is that a particular theorem is provable.

## ———— ———— KEEPING SECRETS ————————

The mathematical basis for such a seemingly impossible scheme is a novel concept called a *zero-knowledge proof*, first formally defined in 1985. The idea, a product of several excitedly interacting groups of computer scientists and mathematicians in the United States, Canada, and Israel, developed quickly. Initially, Shafi Goldwasser, Silvio Micali, and Charles Rackoff, motivated by theoretical questions concerning the efficiency and reliability of computer algorithms, worked out that it was possible to convey that a theorem is proved without having to provide any details of the proof itself. Then Micali, Oded Goldreich, and Avi Wigderson showed that an important class of theorems actually have zero-knowledge proofs. They demonstrated the procedure for a mathematical coloring problem, in which no two points in certain networks of connected points can have the same color. Manuel Blum extended the scheme to cover any mathematical theorem. Amos Fiat and Adi Shamir, taking the idea into the world of spies and sensitive data, used it to suggest a secure, cryptographic method for identifying computer users.

Blum's scheme is interactive. It features a dialog between the prover, who has found a proof for a theorem, and a skeptical verifier. The verifier can ask a special question that requires the equivalent of a yes-or-no answer. If the prover really knows the proof, then he can answer the question correctly every time it is asked. If he doesn't know the proof, the prover has only a 50 percent chance of being right each time. After, say, a dozen tries, the chances of fooling the verifier get very small. Neither the question nor the possible answers give away even a hint of the proof itself—hence, the term zero-knowledge proof.

An example from graph theory shows how the scheme works. Any network of points, or nodes, connected by lines, or edges, is called a graph. In this case, the graph consists of a star-shaped pattern of lines linking 11 points *(see Figure 8.1)*. The prover has found a continuous path along the connecting links that passes only once through each of the 11 points on the graph and returns to where it started. This special type of path is called a Hamiltonian cycle.

Deciding whether a particular graph—a network of points and connecting lines—has a Hamiltonian cycle is often difficult. It involves finding, along the links, a continuous path that passes only once through each of the graph's points and then returns to where it started. For the purposes of a zero-knowledge proof, the star-shaped graph in the example can be redrawn so that all the points fall on the circumference of an imaginary circle, yet the connecting lines still join the same points as in the original graph *(see Figure 8.2)*.

The prover's aim is to persuade a verifier that such a path is known without giving the verifier the slightest idea of how to construct the path. To do this, the prover privately marks 11 nodes along the circumference of a circle and labels them randomly from 1 to 11. Then the nodes on the circle are connected in the same way as the points in the original graph. That is, lines would join nodes 1 and 5, 2 and 6, and so on. Now the resulting diagram is covered up by, say, an erasable opaque film like that used on some lottery or contest tickets.

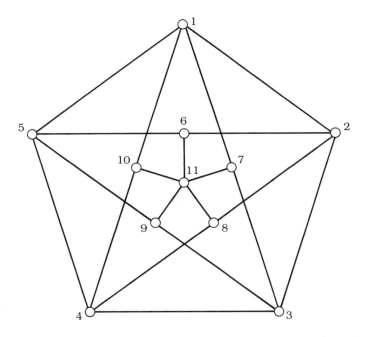

**FIGURE 8.1**   The Hamiltonian cycle in this star-shaped graph passes through points in the order 1, 5, 6, 2, 8, 4, 10, 11, 9, 3, 7, then back to 1.

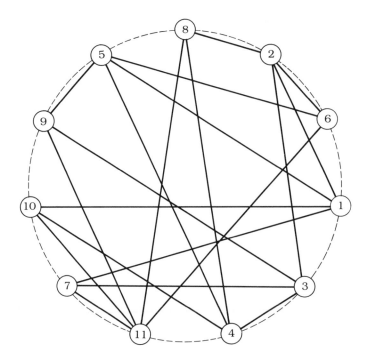

**FIGURE 8.2** A star-shaped graph can be redrawn so that all the points fall on the circumference of an imaginary circle.

The verifier can ask the prover to uncover the complete graph, which shows that all the points are properly linked, or she can ask to see the Hamiltonian cycle. In the latter case, the prover erases enough of the film to reveal only the lines that make up the cycle. He can do this only if he knows the right path. However, because the nodes are still covered up, the verifier doesn't know the actual path from point to point. All she can verify is that a Hamiltonian cycle exists.

The process can be repeated as many times as the verifier wishes. Each time, the prover sets up a new circle diagram, which is then hidden. Because he doesn't know whether the verifier will ask for the graph or the cycle, he has to be ready for either choice and therefore must know the cycle. Failure to produce either the correct graph or the cycle during any turn is equivalent to a wrong answer, and the verifier then knows that either the prover is lying or he doesn't have a correct proof.

Significantly, any mathematical theorem can be converted into a graph in such a way that if the theorem has a proof, then the graph has a Hamiltonian cycle. Such a cycle is easy to find quickly with a proof in hand, and a known cycle can easily be turned into a proof.

But because the graphs may be incredibly complex jumbles of points and lines, simple inspection is rarely enough to pinpoint a given graph's Hamiltonian cycle.

As Blum showed, a zero-knowledge scheme can handle any theorem provable within any logic system. The theorem prover proceeds by converting the proof to a graph with a Hamiltonian cycle, and the rest follows. In the end, all a verifier finds out is that a theorem is provable and that the prover actually knows the proof. And because the prover has to use at least as many words as the proof itself contains, he gives away an upper limit for the proof's length.

As outlined, this scheme sounds somewhat cumbersome. But the opaque film used in the example can be replaced by encryption schemes that hide information. Thus, proof checking can be done electronically, with the whole procedure encoded as strings of binary digits. Such a concept can be used for password protocols and in cryptological games, such as a coin toss by telephone or the exchange of secret keys.

These protocols mean that a computer user entering a password at a computer keyboard would no longer have to worry about the possibility of someone peering over a shoulder and stealing the password as it's typed in. A user would provide the computer with information that persuades the computer that the user knows the password without giving away the password itself. Furthermore, no onlooker would be able to reconstruct the password from information given to the computer. No one could break into the computer system with a purloined password.

## THE BURDEN OF PROOF

Conjecture and proof are the twin pillars of mathematics. Conjecture stands at the cutting edge, in the wild region separating the known from the unknown. Proof stands at the center of mathematics, a solid tower about which an elaborate scaffolding of mathematical theorems is built. The task of mathematicians is to develop conjectures—guesses or hypotheses about mathematical behavior—which they can then attempt to prove. On this basis, mathematics grows, stretching into new fields and revealing hitherto unseen connections between well-established domains.

Like other sciences, mathematics has an experimental component. In the trial-and-error process of developing and proving conjectures, mathematicians collect data and look for patterns and trends. They construct novel forms, seek logical arguments to strengthen a case, and search for counterexamples to destroy an argument or expose an error. In these and other ways, mathematicians act as experimentalists, constantly testing their ideas and methods.

The concept of proof, however, brings something to mathematics that is missing from other sciences. Once the experimental work is done, mathematicians have ways to build a logical argument that pins the label "true" or "false" on practically any conjecture. Physicists can get away with overwhelming evidence to support a theory. In mathematics, a single counterexample is enough to sink a beautiful conjecture. For this reason, too, no set of pretty pictures, collected during countless computer experiments, is complete enough to substitute for a mathematical proof.

Emphasis on proof engenders both a strong conservatism and a deep skepticism among mathematicians. Their professional fame is tied to mathematical proofs of significant conjectures, and the rush to be the first with a correct proof adds considerable confusion. Two recent examples illustrate the pitfalls strewn across a mathematician's path.

In March 1986, British mathematician Colin Rourke and his student Eduardo Rêgo, from Portugal, announced that they had proved the Poincaré conjecture. To mathematicians, especially topologists, proving the Poincaré conjecture was something like being the first to climb Mt. Everest. For more than 80 years, numerous mathematicians had stumbled over this infamous problem, always slipping somewhere along the way. Sometimes, only a tiny gap—a subtle error buried within pages of mathematics—had halted the ascent.

As we saw in Chapter 4, the Poincaré conjecture proposes that any object that mathematically behaves like a three-dimensional sphere is a three-dimensional sphere, no matter how distorted or twisted its shape may be. Although the statement sounds obvious, its difficulty lies in the enumeration of all the different ways in which three-dimensional space can be stretched and molded to form geometric surfaces called manifolds.

Because so many false "proofs" for the Poincaré conjecture had been proferred in the past, mathematicians approached the new claim warily. There was considerable skepticism. Rourke's and Rêgo's approach to proving the conjecture was similar to ideas that had been tried unsuccessfully years before. The biggest problem in verifying the proof, which ran to dozens of pages in manuscript form, was to find mathematicians willing to take the time from their own work. If a mistake happened to be subtle, the task could take years of effort.

Rourke was so sure that he had found the proof that he took an unusual step, for a mathematician, and announced the results in a press release and, later in the year, wrote an article about the proof for a popular science magazine. Meanwhile, most mathematicians keenly interested in the question remained silent and simply waited for better evidence. Some resented the fact that Rourke had publicized his achievement without waiting for verification. Normally, mathematicians quietly pass their manuscripts to their friends and colleagues for comment. Only months or years later, after the mathematical community has applied its stamp of approval, is the work published in a mathematical journal, and the press may then hear of the result.

In November of the same year, Rourke was at the University of California at Berkeley, conducting a seminar to explain and defend his proof. By the end of the week, Rourke's audience, which included some of the world's top topologists, had pointed out a gap in his proof, one that Rourke could not fill. In the end, there was no proof.

The second story has a happier ending. It involves a celebrated conjecture, first stated in 1916 by German mathematician Ludwig Bieberbach, and now known by his name. The Bieberbach conjecture is a statement about the coefficients of power series that represent analytic functions with certain properties. Analytic functions play an important role in calculus and in solving differential equations. These functions of a complex variable can be written as an infinite sum of powers of $z$ having coefficients $a_2$, $a_3$, $a_4$, and so on; in mathematical terms, $f(z) = z + a_2z^2 + a_3z^3 + \ldots$ . The conjecture states that under reasonable restrictions, the coefficients of the power series never get larger than a certain bound.

The conjecture was known to be true only up to the sixth coefficient until Louis de Branges of Purdue University showed that it was true for all coefficients. However, his path to getting his proof of the Bieberbach conjecture recognized was tortuous. In March 1984, de Branges sent out his proof to about a dozen mathematicians for verification. It was part of a 350-page manuscript for a book on power series. All the mathematicians wrote back to say they didn't have the time to read the manuscript. Most doubted that de Branges had succeeded; not only was the manuscript difficult to read, but minor errors were found in the early part of the work.

By coincidence, de Branges was scheduled to go to the University of Leningrad in June. In his proof, de Branges had proved a conjecture first proposed by I. M. Milin, who happened to be in Leningrad at the time. Milin's conjecture directly implies the truth of the Bieberbach conjecture. In Leningrad, de Branges explained his proof in detail, and his hosts did the necessary verification, boiling the proof down to its essentials. As news of the proof spread, other mathematicians were able to examine the proof closely and to confirm its validity.

One other remarkable coincidence played an important role in establishing de Branges' proof. The proof depended on an inequality concerning certain special functions. Working with Walter Gautschi, de Branges was able to test the validity of the inequality up to thirtieth coefficient by using a supercomputer. The results were encouraging but did not provide a valid proof. Gautschi and de Branges then found out that two mathematicians, Richard Askey and George Gasper, had established the validity of the inequality in an entirely different context several years before. With that piece of evidence, de Branges' proof was complete.

De Branges' proof has been streamlined so that it now looks easy. What the simplified version hides, however, is any indication why anyone would ever have bothered to take the particular route that de Branges did to proving the Bieberbach conjecture.

The case of the Bieberbach conjecture demonstrates one other important lesson. Although the final proof of an important conjecture may be a great achievement, it generally rests on the work—the ideas, conjectures, and techniques—of many mathematicians. The communal building process emphasizes again the importance of proofs for establishing a firm foundation on which future generations of mathematicians can build.

The mathematical community also likes to keep its disputes within the family and sweep messy details away. It's not surprising that outsiders rarely see mathematics as the exciting, human endeavor it is. What emerges instead is an elegant, remote architecture, with the scaffolding down and the blueprints stored away. The human element is hidden.

Mathematics is an exciting, dynamic field that continues to generate provocative ideas and novel concepts. Some notions quickly find their way into applications; others rest on their own intrinsic beauty; still others occupy a modest place in the growing structure of mathematics itself. As mathematician Rick Norwood eloquently noted in his essay "In Abstract Terrain,"

So it goes, new mathematics from old, curving back, folding and unfolding, old ideas in new guises, new theorems illuminating old problems. Doing mathematics is like wandering through a new countryside. We see a beautiful valley below us, but the way down is too steep, and so we take another path, which leads us far afield, until, by a sudden and unexpected turning, we find ourselves walking in the valley.

# FURTHER READING

## 1 EXPLORATIONS

*Mathematical Sciences: A Unifying and Dynamic Resource.* Panel on Mathematical Sciences, National Research Council. National Academy Press, 1986.

*Mathematical People.* Donald J. Albers and G. L. Alexanderson, editors. Birkhauser, 1985.

"The Solution of the Four-Color-Map Problem," Kenneth Appel and Wolfgang Haken in *Scientific American,* Vol. 237, October 1977, pages 108–121.

"The Four Color Proof Suffices," K. Appel and W. Haken in *The Mathematical Intelligencer,* Vol. 8, No. 1, 1986, pages 10–20.

*Mathematics Today.* Lynn Arthur Steen, editor. Springer-Verlag, 1978.

*The Problems of Mathematics.* Ian Stewart. Oxford University Press, 1987.

## 2 PRIME PURSUITS

*Number Theory in Science and Communication.* M. R. Schroeder. Springer-Verlag, 1984.

"The Invincible Primes," Solomon W. Golomb in *The Sciences,* Vol. 25, No. 2, March-April 1985, pages 50–57.

"Recent Developments in Primality Testing," Carl Pomerance in *The Mathematical Intelligencer,* Vol. 3, No. 2, 1981, pages 97–105.

"Factoring on a Computer," H. C. Williams in *The Mathematical Intelligencer,* Vol. 6, No. 3, 1984, pages 29–36.

"Factoring Large Numbers on a Pocket Calculator," W. D. Blair, C. B. LaCampagne, and J. L. Selfridge in *Mathematics Magazine,* Vol. 59, No. 5, December 1986, pages 802–803.

"The Remarkable Lore of the Prime Numbers," Martin Gardner in *Scientific American.* Vol. 210, March 1964, pages 120–128.

# 3 TWISTS OF SPACE

*Mathematics and Optimal Form.* Stefan Hildebrandt and Anthony Tromba. Scientific American Books, 1985.

"Soap Films and Problems without Unique Solutions," Frank Morgan in *American Scientist,* Vol. 74, May–June 1986, pages 232–235.

"The Computer-Aided Discovery of New Embedded Minimal Surfaces," David Hoffman in *The Mathematical Intelligencer,* Vol. 9, No. 3, 1986, pages 8–21.

"The Geometry of Soap Films and Soap Bubbles," Frederick J. Almgren, Jr., and Jean E. Taylor in *Scientific American,* Vol. 235, July 1976, pages 82–93.

"Catalog of Saddle Shaped Surfaces in Crystals," J. E. Taylor and J. W. Cahn in *Acta Metallurgica,* Vol. 34, No. 1, 1986, pages 1–12.

"The Theory of Knots," Lee Neuwirth in *Scientific American,* Vol. 240, June 1979, pages 110–124.

# 4 SHADOWS FROM HIGHER DIMENSIONS

*Flatland.* Edwin A. Abbott. Dover Publications, Inc., 1952.

*The Fourth Dimension and Non-Euclidean Geometry in Art.* Linda D. Henderson. Princeton University Press, 1983.

"Computer Graphics and the Geometry of S3," Huseyin Koçak and David Laidlaw in *The Mathematical Intelligencer,* Vol. 9, No. 1, 1987, pages 8–10.

"The Mathematics of Three-Dimensional Manifolds," William P. Thurston and Jeffrey R. Weeks in *Scientific American,* Vol. 251, July 1984, pages 108–120.

*A Topological Picturebook.* George K. Francis. Springer-Verlag, 1987.

"The Allocation of Resources by Linear Programming," Robert G. Bland in *Scientific American,* Vol. 244, June 1981, pages 126–144.

"Karmarkar's Algorithm and Its Place in Applied Mathematics," Gilbert Strang in *The Mathematical Intelligencer,* Vol. 9, No. 2, 1987, pages 4–10.

# 5  ANTS IN LABYRINTHS

*The Fractal Geometry of Nature.* Benoit B. Mandelbrot. W. H. Freeman, 1982.

"Of Fractal Mountains, Graftal Plants and Other Computer Graphics at Pixar," A. K. Dewdney in *Scientific American,* Vol. 255, December 1986, pages 14–20.

"Solution of an Inverse Problem for Fractals and Other Sets," M. F. Barnsley, V. Ervin, D. Hardin, and J. Lancaster in *Proceedings of the National Academy of Sciences* (USA), Vol. 83, April 1986, pages 1975–1977.

*On Growth and Form.* H. Eugene Stanley and Nicole Ostrowsky, editors. Martinus Nijhoff Publishers, 1986.

# 6  THE DRAGONS OF CHAOS

"Classical Chaos," Roderick V. Jensen in *American Scientist,* Vol. 75, March–April 1987, pages 168–181.

"Order in Chaos," John H. Hubbard in *Engineering: Cornell Quarterly,* Vol. 20, No. 3, Winter 1986, pages 20–26.

"Chaos," James P. Crutchfield, J. Doyne Farmer, Norman H. Packard, and Robert S. Shaw in *Scientific American,* Vol. 255, December 1986, pages 46–57.

"Roads to Chaos," Leo P. Kadanoff in *Physics Today,* Vol. 36, December 1983, pages 46–53.

*The Beauty of Fractals.* H.-O. Peitgen and P. H. Richter. Springer-Verlag, 1986.

*Chaos: The Making of a New Science.* James Gleick. Viking, 1987.

"How Random Is a Coin Toss?" Joseph Ford in *Physics Today,* Vol. 36, April 1983, pages 40–47.

"Chaotic Bursts in Nonlinear Dynamical Systems," Robert L. Devaney in *Science,* Vol. 235, 16 January 1987, pages 342–345.

"Strange Attractors," David Ruelle in *The Mathematical Intelligencer,* Vol. 3, No. 3, 1980, pages 126–137.

"Strange Attractors: Mathematical Patterns Delicately Poised Between Order and Chaos," Douglas R. Hofstadter in *Scientific American,* Vol. 245, November 1981, pages 22–43.

# 7 LIFE STORIES

*The Recursive Universe.* William Poundstone. William Morrow, 1985.

*Winning Ways for Your Mathematical Plays.* Elwyn R. Berlekamp, John H. Conway, and Richard K. Guy. Academic Press, 1982.

"Computer Software in Science and Mathematics," Stephen Wolfram in *Scientific American,* Vol. 251, September 1984, pages 188–203.

"Building Computers in One Dimension Sheds Light on Irreducibly Complicated Phenomena," A. K. Dewdney in *Scientific American,* Vol. 252, May 1985, pages 18–30.

*Tilings and Patterns.* Branko Grünbaum and G. C. Shephard. W. H. Freeman, 1987.

"Quasicrystals," Paul Joseph Steinhardt in *American Scientist,* Vol. 74, November–December 1986, pages 586–597.

"Quasicrystals," David R. Nelson in *Scientific American,* Vol. 255, August 1986, pages 42–51.

"Extraordinary Nonperiodic Tiling That Enriches the Theory of Tiles," Martin Gardner in *Scientific American,* Vol. 236, January 1977, pages 110–122.

"Pentaplexity," Roger Penrose in *Geometrical Combinatorics.* F. C. Holroyd and R. J. Wilson, editors. Pitman, 1984.

"Cellular Automata as Models of Complexity," Stephen Wolfram in *Nature,* Vol. 311, 4 October 1984, pages 419–424.

*Wheels, Life, and Other Mathematical Amusements.* Martin Gardner. W. H. Freeman, 1983.

# 8 IN ABSTRACT TERRAIN

"Poincaré's Perplexing Problem," Colin Rourke and Ian Stewart in *New Scientist,* Vol. 111, 4 September 1986, pages 41–45.

"The Bieberbach Conjecture," Paul Zorn in *Mathematics Magazine,* Vol. 59, No. 3, June 1986, pages 131–148.

"In Abstract Terrain," Rick Norwood in *The Sciences,* Vol. 22, No. 9, December 1982, pages 13–18.

# SOURCES
# OF
# ILLUSTRATIONS

**P A G E**  3 (Figure 1.1, *bottom*)
Created by Edward F. Moore, University of Wisconsin (*Scientific American,* October 1977, page 109)

**P A G E**  6
Charles S. Peskin, New York University

**P A G E**  22
*Scientific American,* March 1964, page 120

**P A G E**  24
Los Alamos National Laboratory, Los Alamos, New Mexico (*Scientific American,* March 1964, page 122)

**P A G E**  33
*Scientific American,* December 1982, page 144

**P A G E**  47
Frei Otto, Institut für leichte Flächentragwerke, Stuttgart

**P A G E S**  49, 50, 51 (Figure 3.5, *bottom*), 52 (Figure 3.6, *top*), 54, 55, and 56
Stefan Hildebrandt and Anthony Tromba, *Mathematics and Optimal Form,* © 1985 by Scientific American Books, Inc.

**P A G E S**  59 and 60
J. T. Hoffman and D. Hoffman, University of Massachusetts

**P A G E**  64 (Figure 3.13, *left*)
Adapted from photo courtesy of R. J. Gray (retired), Oak Ridge National Laboratory

**P A G E**  64 (Figure 3.13, *right*)
Cyril Stanley Smith, *A Search for Structure,* The MIT Press, © 1981 by The Massachusetts Institute of Technology

**P A G E S**  65, 67, 68, and 69 (Figure 3.17, *left*)
Jean E. Taylor, Rutgers University

**P A G E**  69 (Figure 3.17, *right*)
John L. Walter, Research and Development Center, General Electric Company

**P A G E  71**
*Scientific American*, June 1979, page 112

**P A G E  72**
Joan S. Birman, Columbia University

**P A G E  73** (Figure 3.20)
Kenneth C. Millett, University of California at Santa Barbara

**P A G E S  79 and 80**
Nicholas R. Cozzarelli, University of California at Berkeley and Steven A. Wasserman, University of Texas at Austin

**P A G E  86**
*Scientific American*, June 1987, page 112

**P A G E S  89** (Figure 4.4) and **95**
*Scientific American*, April 1986, page 20

**P A G E  90**
Thomas F. Banchoff, Brown University

**P A G E  93**
Lynn Arthur Steen (editor), *Mathematics Today*, Springer-Verlag, © 1978 by the Conference Board of the Mathematical Sciences

**P A G E  96**
Thomas F. Banchoff, Hüseyin Koçak, and David Laidlaw, Brown University

**P A G E  98**
David Laidlaw and Hüseyin Koçak, Brown University

**P A G E  102**
This first appeared in *New Scientist*, London, the weekly review of science and technology

**P A G E  110**
*Scientific American*, June 1981, page 127

**P A G E S  117, 120, 121** (Figure 5.3), **126, 133, 136, 157, and 160**
Benoit B. Mandelbrot, *The Fractal Geometry of Nature*, W. H. Freeman and Company, © 1982 by Benoit B. Mandelbrot

**P A G E**   121 (Figure 5.4)
Leonard M. Blumenthal and Karl Menger, *Studies in Geometry,* © 1970 by
W. H. Freeman and Company

**P A G E**   125
Alvy Ray Smith, Lucasfilm Ltd. (*Scientific American,* September 1984, page
156)

**P A G E**   127
Peter Oppenheimer, New York Institute of Technology

**P A G E S**   129 and 132
Michael F. Barnsley, Georgia Institute of Technology

**P A G E S**   138 and 140
Paul Meakin, Du Pont Experimental Station

**P A G E**   141 (Figure 5.14, *top*)
Roy Richter, GM Research Laboratories (*Scientific American,* January 1987,
page 98)

**P A G E**   141 (Figure 5.14, *bottom*)
Nancy Hecker and David G. Grier, University of Michigan (*Scientific American,* January 1987, page 98)

**P A G E**   147
Edward N. Lorenz, Massachusetts Institute of Technology

**P A G E**   152
*Scientific American,* July 1987, page 110

**P A G E**   154
John Guckenheimer, Cornell University

**P A G E S**   159, 161, 163, and 168
Reprinted with permission from *The Beauty of Fractals* by Heinz-Otto Peitgen and Peter H. Richter, Springer-Verlag, 1986

**P A G E**   169
S. Burns, H. Benzinger, and J. Palmore, University of Illinois

**P A G E**   173
Robert L. Devaney, Boston University

**P A G E**   178 (Figure 7.2)
*Scientific American,* November 1984, page 40

**P A G E   178** (Figure 7.3)
*Abacus,* © 1987 by Springer-Verlag

**P A G E S   182, 184, 185, 186 and 187**
Stephen Wolfram, University of Illinois

**P A G E S   190, 192 and 193**
Rick Durrett, Cornell University

**P A G E   198**
*Scientific American,* August 1984, page 20

**P A G E S   202 and 204**
*Scientific American,* January 1977, pages 115 and 116

**P A G E   203**
*The Mathematical Intelligencer,* © 1979 by Springer-Verlag

**P A G E   205** (Figure 7.21)
Seymour Haber, National Bureau of Standards

**P A G E   208**
Michel Duneau and André Katz, École Polytechnique, Palaiseau Cedex, France

**P A G E   209**
*Scientific American,* August 1986, page 48

**P A G E   210** (Figure 7.25)
Howland A. Fowler, National Bureau of Standards

**C O L O R   P L A T E   1**
Minimal Möbius band, discovered by W. H. Meeks; computer-generated image © 1986 by J. T. Hoffman, University of Massachusetts

**C O L O R   P L A T E   2**
Genus-one, Costa-Hoffman-Meeks embedded minimal surface; computer-generated image © 1987 by J. T. Hoffman and D. Hoffman, University of Massachusetts

**C O L O R   P L A T E   3**
The core of the four-lobed Wente torus; computed by J. Spruck, A. Eydeland, and M. Callahan; image © 1987 by J. T. Hoffman, University of Massachusetts

**C O L O R   P L A T E   4**
Produced at Brown University by Hüseyin Koçak and David Laidlaw, in collaboration with T. Banchoff, F. Bisshopp, and D. Margolis

228

COLOR PLATE 5
Richard F. Voss, IBM Research

COLOR PLATE 6
Michael F. Barnsley, © 1987 Georgia Tech Research Corp.

COLOR PLATE 7
Paul Meakin, Du Pont Experimental Station

COLOR PLATE 8
James P. Crutchfield, University of California at Berkeley (*Scientific American*, December 1986, page 57)

COLOR PLATE 9
Generated on NASA's Massively Parallel Processor by Edward Seiler, NASA Goddard Space Flight Center

COLOR PLATES 10 and 11
S. Burns, H. Benzinger, and J. Palmore, University of Illinois

COLOR PLATE 12
Robert L. Devaney, Boston University

COLOR PLATE 13
Stephen Wolfram, University of Illinois, (*Scientific American*, September 1984, page 199)

COLOR PLATE 14
Norman H. Packard, University of Illinois (*Scientific American*, September 1984, page 189)

COLOR PLATE 15
Richard Durrett, Cornell University

COLOR PLATE 16
Paul J. Steinhardt, University of Pennsylvania

# INDEX